CONTROL THEORY IN MATHEMATICAL ECONOMICS

PURE AND APPLIED MATHEMATICS

A Program of Monographs, Textbooks, and Lecture Notes

Contributions to *Lecture Notes in Pure and Applied Mathematics* are reproduced by direct photography of the author's typewritten manuscript. Potential authors are advised to submit preliminary manuscripts for review purposes. After acceptance, the author is responsible for preparing the final manuscript in camera-ready form, suitable for direct reproduction. Marcel Dekker, Inc. will furnish instructions to authors and special typing paper. Sample pages are reviewed and returned with our suggestions to assure quality control and the most attractive rendering of your manuscript. The publisher will also be happy to supervise and assist in all stages of the preparation of your camera-ready manuscript.

LECTURE NOTES
IN PURE AND APPLIED MATHEMATICS

1. *N. Jacobson,* Exceptional Lie Algebras
2. *L.-Å. Lindahl and F. Poulsen,* Thin Sets in Harmonic Analysis
3. *I. Satake,* Classification Theory of Semi-Simple Algebraic Groups
4. *F. Hirzebruch, W. D. Newmann, and S. S. Koh,* Differentiable Manifolds and Quadratic Forms
5. *I. Chavel,* Riemannian Symmetric Spaces of Rank One
6. *R. B. Burckel,* Characterization of C(X) Among Its Subalgebras
7. *B. R. McDonald, A. R. Magid, and K. C. Smith,* Ring Theory: Proceedings of the Oklahoma Conference
8. *Y.-T. Siu,* Techniques of Extension of Analytic Objects
9. *S. R. Caradus, W. E. Pfaffenberger, and B. Yood,* Calkin Algebras and Algebras of Operators on Banach Spaces
10. *E. O. Roxin, P.-T. Liu, and R. L. Sternberg,* Differential Games and Control Theory
11. *M. Orzech and C. Small,* The Brauer Group of Commutative Rings
12. *S. Thomeier,* Topology and Its Applications
13. *J. M. López and K. A. Ross,* Sidon Sets
14. *W. W. Comfort and S. Negrepontis,* Continuous Pseudometrics
15. *K. McKennon and J. M. Robertson,* Locally Convex Spaces
16. *M. Carmeli and S. Malin,* Representations of the Rotation and Lorentz Groups: An Introduction
17. *G. B. Seligman,* Rational Methods in Lie Algebras
18. *D. G. de Figueiredo,* Functional Analysis: Proceedings of the Brazilian Mathematical Society Symposium
19. *L. Cesari, R. Kannan, and J. D. Schuur,* Nonlinear Functional Analysis and Differential Equations: Proceedings of the Michigan State University Conference
20. *J. J. Schäffer,* Geometry of Spheres in Normed Spaces
21. *K. Yano and M. Kon,* Anti-Invariant Submanifolds
22. *W. V. Vasconcelos,* The Rings of Dimension Two
23. *R. E. Chandler,* Hausdorff Compactifications
24. *S. P. Franklin and B. V. S. Thomas,* Topology: Proceedings of the Memphis State University Conference
25. *S. K. Jain,* Ring Theory: Proceedings of the Ohio University Conference
26. *B. R. McDonald and R. A. Morris,* Ring Theory II: Proceedings of the Second Oklahoma Conference
27. *R. B. Mura and A. Rhemtulla,* Orderable Groups
28. *J. R. Graef,* Stability of Dynamical Systems: Theory and Applications
29. *H.-C. Wang,* Homogeneous Banach Algebras
30. *E. O. Roxin, P.-T. Liu, and R. L. Sternberg,* Differential Games and Control Theory II
31. *R. D. Porter,* Introduction to Fibre Bundles
32. *M. Altman,* Contractors and Contractor Directions Theory and Applications
33. *J. S. Golan,* Decomposition and Dimension in Module Categories
34. *G. Fairweather,* Finite Element Galerkin Methods for Differential Equations
35. *J. D. Sally,* Numbers of Generators of Ideals in Local Rings
36. *S. S. Miller,* Complex Analysis: Proceedings of the S.U.N.Y. Brockport Conference
37. *R. Gordon,* Representation Theory of Algebras: Proceedings of the Philadelphia Conference
38. *M. Goto and F. D. Grosshans,* Semisimple Lie Algebras
39. *A. I. Arruda, N. C. A. da Costa, and R. Chuaqui,* Mathematical Logic: Proceedings of the First Brazilian Conference
40. *F. Van Oystaeyen,* Ring Theory: Proceedings of the 1977 Antwerp Conference

41. *F. Van Oystaeyen and A. Verschoren,* Reflectors and Localization: Application to Sheaf Theory
42. *M. Satyanarayana,* Positively Ordered Semigroups
43. *D. L. Russell,* Mathematics of Finite-Dimensional Control Systems
44. *P.-T. Liu and E. Roxin,* Differential Games and Control Theory III: Proceedings of the Third Kingston Conference, Part A
45. *A. Geramita and J. Seberry,* Orthogonal Design: Quadratic Forms and Hadamard Matrices
46. *J. Cigler, V. Losert, and P. Michor,* Banach Modules and Functors on Categories of Banach Spaces
47. *P.-T. Liu and J. G. Sutinen,* Control Theory in Mathematical Economics: Proceedings of the Third Kingston Conference, Part B

Other Volumes in Preparation

CONTROL THEORY IN MATHEMATICAL ECONOMICS

Proceedings of the Third Kingston Conference, Part B

Edited by

PAN-TAI LIU
Department of Mathematics
University of Rhode Island
Kingston, Rhode Island

JON G. SUTINEN
Department of Resource Economics
University of Rhode Island
Kingston, Rhode Island

MARCEL DEKKER, INC. New York and Basel

Library of Congress Cataloging in Publication Data

Kingston Conference on Differential Games and Control
Theory, 3d, University of Rhode Island, 1978.
Control theory in mathematical economics.

(Lecture notes in pure and applied mathematics ; v. 47)
1. Control theory--Congresses. 2. Economics, Mathematical--Congresses. I. Liu, Pan-Tai. II. Sutinen, Jon G. III. Title.
HB135.K53 1978 330'.01'51 79-13785
ISBN 0-8247-6852-3

MARCEL DEKKER, INC.
270 Madison Avenue, New York, New York 10016

Current printing (last digit)
10 9 8 7 6 5 4 3 2 1

PRINTED IN THE UNITED STATES OF AMERICA

PREFACE

In recent years, mathematical theory and techniques in the control of dynamic systems have emerged as a useful discipline for the study of many economic and management science problems. The papers appearing in this volume cover a wide range of mathematical applications, including the pure theory of production, exploitation of a fishery resource, advertising, etc.

Some of the papers are characterized by their novel approaches to the problems, while others contain significant results that represent some of the recent developments in a special area.

These papers were presented at the Third Kingston Conference on Differential Games and Control Theory with a theme on dynamic optimization and mathematical economics in June 1978 at Kingston, Rhode Island. The conference was sponsored by the Office of Naval Research and the University of Rhode Island. We'd like to thank, in particular, Dr. Neal Glassman and Dr. Thomas Varley of ONR and Dr. William Ferrante, Vice President of URI, for their support. Finally, we'd also like to thank Mrs. Marguerite Ellis for her excellent typing of the manuscript.

Pan-Tai Liu and Jon G. Sutinen

LIST OF CONTRIBUTORS

HOWARD BLUM, Rutgers University, Newark, New Jersey

J. CASE, Johns-Hopkins University, Baltimore, Maryland

SIMONE CLEMHOUT, Cornell University, Ithaca, New York

K. DEAL, McMaster University, Hamilton, Ontario, Canada

M. H. I. DORE, University of Saskatchewan, Saskatoon, Canada

KUAN-PIN LIN, State University of New York at Stony Brook, Stony Brook, New York, and Harvard University, Cambridge, Massachusetts

LEONARD J. MIRMAN, University of Illinois, Urbana-Champaign, Illinois

S. P. SETHI, University of Toronto, Toronto, Ontario, Canada

H. M. SHEFRIN, University of Rochester, Rochester, New York, and University of Santa Clara, Santa Clara, California

GILBERT S. SUZAWA, University of Rhode Island, Kingston, Rhode Island

G. L. THOMPSON, Carnegie-Mellon University, Pittsburgh, Pennsylvania

C. WRATHER, University of Texas, Austin, Texas

P. L. YU, University of Kansas, Lawrence, Kansas

CONTENTS

CONTROL THEORY IN MATHEMATICAL ECONOMICS

A THEORY OF LONG-RUN PRODUCTION FUNCTION

Gilbert S. Suzawa

University of Rhode Island
Kingston, Rhode Island

§1. INTRODUCTION

Since R. W. Shepard [5] first introduced the idea of the "homothetic" property to the theory of the production function, these production functions have undergone intensive investigations and new forms are increasingly being applied to study production behavior. Homotheticity is a geometric property. In terms of production functions, this property is attributed to the two-dimensional input space. Consequently, homothetic production functions are more properly designated as "homothetic isoquant" production functions as in Clemhout [1]. In general, however, homothetic production functions imply classes of specific production functions which are related by means of an order preserving transformation which preserves the homothetic property of the isoquant map.

An advantage of the homotheticity property is that it enables one to isolate the elasticity of factor substitution properties of production from variations in the output level. Since homothetic production functions are related by monotone increasing transformations, the shape of the isoquant map is unaltered. Hence, whether the underlying isoquants manifest constant or variable elasticities of substitution, this characteristic will be preserved for the entire class of functions. In other words, a class of homothetic production functions differ only in that there is an order preserving renumbering of the isoquant map.

It is well known that homogeneous functions are a subset of homothetic functions. There is no reason to characterize the properties of the isoquant map of a class of homothetic production functions in terms of a homogeneous function, unless of course there is some empirical basis to do so. In studies such as Clemhout [1], Zellner and Revankar [6], and Meyer [2], the practice has been to specify the two-dimensional properties of the input surface in terms of a homogeneous function. Clemhout, for example, starts by assuming the general homothetic form

$$Y = F[f(K,L)] \tag{i}$$

where F is a monotone increasing transformation and f(K,L) is homogeneous of degree one in capital and labor.

Recently, Sato, Koizumi, and Wolkowitz [4] have introduced homothetic production functions based on nonhomogeneous intermediate functions. Their approach to homothetic production function goes essentially as follows. Suppose that the marginal rate of substitution between capital and labor is some function of only these inputs:

$$-\frac{dK}{dL} = \phi(K,L) \tag{ii}$$

If the ϕ-function is a single-valued increasing function of the capital/labor ratio, then the isoquant map is defined by a homogeneous function.[1] However, there is no *a priori* reason to presume this to always be true. In principle, (ii) can be a differential equation of any order. Its solution may be derived by finding an integrating factor which transforms it into an exact differential equation. Let u be an integrating factor such that

$$udK + u\phi(K,L) = 0 \qquad \text{(iii)}$$

Differential equation (iii) can thus be integrated to obtain

$$Z = f(K,L) \qquad \text{(iv)}$$

where Z is the constant of integration term. However, since the integrating factor is generally not unique, a more general factor, such as uF(Z), will yield a solution

$$Y = F(Z) = F[f(K,L)] \qquad \text{(v)}$$

One of the widely recognized applications of homothetic production functions is the study of economies of scale in production. The monotone increasing transformations which generate a family of homothetic production functions also determine the contour of the output surface of each individual function--each transformation describes the manner in which the homothetic isoquant map is renumbered. Thus, the curvature of the output surface, as exemplified by the numbering of the isoquant map, is a representation of the standard neoclassical conception of returns to scale. By finding the output surface which corresponds best to observations, the scale behavior of production is presumably isolated.

---

[1]An implication of this approach to homothetic production functions is that the isoquant map need not have constant marginal rates of substitution along a given ray from the origin.

In applying homothetic production functions to estimate the scale aspect of production, there is the problem of obtaining "neutral" returns to scale measures. The homotheticity property of isoquants allows the elasticity of factor substitution to be free of the output level. However, unless the isoquants are derived under conditions of constant returns to scale, the output surface will not be free of influences associated with input variations. Consequently, the standard practice in empirical studies has been to derive homothetic isoquant relationships under the condition of some constant returns to scale intermediate function. However, that is not the only procedure that can be marshalled to obtain neutral returns to scale measures. Zellner and Revankar [6] have suggested that returns to scale be defined in terms of the elasticity of the F-transformation; i.e., $\frac{dY}{dZ}\frac{Z}{Y}$. Therefore, even if the intermediate function is not homogeneous of degree one in terms of the inputs, any nonneutral effects should be reflected in the intermediate output value and factored out when the elasticity is computed.

Although homothetic production functions provide a convenient apparatus to measure scale effects, we have a limited understanding of the various factors that could influence the scale features of homothetic production functions. Zellner and Revankar have proposed that the elasticity of the F-transformation is a function of the output level. Their hypothesis is interesting but not novel. Furthermore, they offer no explanation as to *how* the output level determines the nature of the F-transformation. They merely rely on *ad hoc* "preassigned" returns to scale functions for empirical and pedagogical purposes.

Homothetic production functions provide a useful framework to study the evolution of production processes over time.[2] The transformations which relate isoquants in a homothetic manner could be made dependent on some systematic long-run influence. Conceived in this fashion, the F-transformation would characterize the manner in which the output surface changes over time.

In this paper, a model of long-run managerial effort optimization is formulated and solved. By directing managerial effort toward motivating workers, the managerial effort input enables measured output to diverge from "intermediate" output. A constraint on the allocation of managerial effort capacity between motivation and innovation is summarized by a differential equation in which the time rate of change of intermediate output is a function of both measured and intermediate outputs. By postulating that managers seek to maximize their discounted utility based on profits and by applying the Pontryagin Maximum Principle, we derive a class of homothetic production functions. As special cases, the implications of isoelastic and quadratic utility functions, in conjunction with linear profit and linear dynamic constraint functions are examined in detail.

§2. MODEL OF MANAGERIAL EFFORT OPTIMIZATION

In order to emphasize the role of systematic managerial behavior in determining the profile of the firm's long-run production function and to satisfy the constraints of the mathematical technique employed, some heroic assumptions are made. The most unrealistic of these assumptions is that the firm's

---

[2] I am in debt to Professor Ryuzo Sato for the idea of using the framework of homothetic production functions to develop a theory of long-run production behavior.

physical inputs do not grow. The impetus for the long-run growth in the firm's real output is assumed to be induced disembodied technological change. Thus, the dynamic model of the firm developed in this and the next section must be construed as a special case of a more general growth model.

§2.1 Dynamic Constraint

We conceive M to be our manager's effort capacity. Managerial effort can be directed toward several types of functions generally essential to the operation of a simple firm. For example, a manager could direct his effort toward allocation of resources such as labor activity, material flows and plant and equipment utilization, or he can search for new information (research) and utilize this new information (development), or he may devote his effort to external matters. When a manager's effort is directed toward a certain type of function, this managerial effort input makes it possible for the firm's actual output to diverge from its "intermediate" output, where we conceive the intermediate output to be the maximum output obtainable from physically measured inputs. This divergence between actual output and intermediate output is possible because one class of the firm's nonmanagerial inputs is derived from human sources, i.e., hired laborers. Being human-based, there are really two dimensions to this particular type of input. One dimension is time and the other is the work effort per unit of time, or the "effort rate." There are some grounds to believe that if a manager were to divert some of his efforts to a certain type of activity, then the output generated by his firm would increase with no change in the work force, rate of capital utilization, and no improvement in allocative efficiency. These activities accomplish what we shall call the "motivational" function. The effort expended in performing the motivational function is designated "motivational effort." Since the manager's effort capacity is finite, we assume that an increase in effort directed toward performing the motivational function implies a reduction in research and development (R&D) effort.

Notations

M_1 Managerial effort directed toward motivational function

M_2 Managerial effort directed toward research and development

θ Proportion of the manager's total effort capacity devoted to motivational effort; i.e., $\theta = M_1/M$, where $0 \leqq \theta \leqq 1$

X Vector of nonmanagerial inputs measured in physical terms

Y Firm's actual output

Z Firm's intermediate output

T State of disembodied technology

The firm's short-run production function is characterized by

$$Y = F(M_1, Z) \tag{1}$$

Since we assume that the manager is on his effort frontier, function (1) defines the maximum firm output obtainable from both physical and "motivated" human inputs. We assume that the marginal product of motivational managerial effort is positive, $\partial F/\partial M_1 < 0$, and diminishing, $\partial^2 F/\partial M_1{}^2 < 0$. We also assume $\partial F/\partial Z > 0$, but make no presumptions about $\partial^2 F/\partial Z^2$.

The intermediate output function is assumed to be

$$Z = Tf(X) \tag{2}$$

Although efficient allocation of the nonmanagerial resources requires managerial effort, we have disregarded this aspect of managerial effort allocation. This is not an unreasonable assumption. If the manager desires to increase current profit, diverting his effort from the allocative function to the motivational function would be a more conflicting policy that diverting some of his effort from the R&D function. Thus, we must view the motivational and allocative functions as complementary. To avoid even dealing with complementary functions, we assume that given T, Z is the maximum output that can be obtained from physically measured nonmanagerial inputs. The variable T is an index of disembodied Hicks neutral technology. We assume that any technological change that occurs affects only the maximum output obtainable from the physical inputs.

Taking the differential of (2) with respect to time, we get

$$\dot{Z} = \dot{T}f(X) + Tf'(X)\dot{X} \tag{3}$$

We assume physical inputs to be fixed.

$$X = \bar{X} \tag{4}$$

This implies that any change in intermediate output over time is assumed to be due only to disembodied technological progress.

We assume the following technological progress function:

$$\dot{T} = g(Z, M_2) \tag{5}$$

Technological change is a function of the level of intermediate output and of managerial research and development effort. The intermediate output level Z is indicative of the past accumulation of technical knowledge. The larger Z is, the greater the increase in $\dot{T}$ will be; i.e., $\partial g/\partial Z > 0$. This is not unreasonable since the greater the past accumulation of basic principles, or R&D methodology, and of research and engineering skills, the greater will be the capacity to enhance technical knowledge. However, there presumably exists diminishing returns to the past's contribution to the advancement of technical knowledge without infusion of current managerial R&D effort, $\partial^2 g/\partial Z^2 < 0$. The current application of managerial effort (M_2) toward R&D also tends to increase technical knowledge; i.e., $\partial g/\partial M_2 > 0$. However, as in the case of motivational managerial efforts, there exists diminishing returns to managerial efforts directed toward R&D activities, i.e., $\partial^2 g/\partial M_2{}^2 < 0$.

The manager's effort capacity is assumed to be constant over time. Let $\theta = M_1/\bar{M}$, $0 \leq \theta \leq 1$. Since $\bar{M} = M_1 + M_2$, $M_1 = \theta\bar{M}$, and $M_2 = (1 - \theta)\bar{M}$. By substitution, we have

$$Y = F(\theta\bar{M}, Z) \tag{1a}$$

and

$$\dot{Z} = g[Z_1(1 - \theta)\bar{M}]f(\bar{X}) \tag{3a}$$

Variations in θ reflect the allocation of capacity managerial effort between the motivational function and the R&D

function. This allocation process has an effect on current firm output and the firm's future intermediate output. By increasing θ, Y will increase. However, an increase in θ implies a decrease in time rate of increase in intermediate output, i.e., $\partial \dot{Z}/\partial \theta < 0$.

For convenience, the dynamic constraint placed on allocation of capacity managerial effort can be characterized by the following function:

$$\dot{Z} = \psi(Y,Z) \tag{6}$$

with the following properties[3]: $\psi_Y < 0$, $\psi_{YY} < 0$, $\psi_Z > 0$, and $\psi_{ZZ} < 0$. In the optimization problem to be formulated next, we shall deal explicitly only with θ.

§2.2 Optimal Control Problem

There is no consensus in the literature on the theory of the firm regarding the relevant behavioral assumptions of the firm's decision-makers. The traditional neoclassical assumption was that entrepreneur-managers maximized profit. More recently, economists adhering to the managerial enterprise position emphasized the assumption that managers have a high degree of discretion with regard to setting the goal of a firm and they seek to maximize some sort of utility function in which profit may be an argument, but not the only one. We shall assume that our hypothetical manager seeks to maximize

---

[3]Where feasible, we shall use subscripts to denote partial derivatives and primes to denote derivatives. The dot (·) over a variable, of course, denotes a time derivative.

a time independent utility function defined for profit and possibly some other autonomous variables.

$$U = U(\pi) \tag{7}$$

We assume $U' > 0$ and $U'' < 0$.

The firm's real profit is defined as

$$\pi = Y - \beta Z \tag{8}$$

This definition assumes that managerial effort is not explicitly priced or is imperfectly priced. The value of nonpriced managerial effort is reflected in profit, the residual income. The physical inputs are assumed to be paid a money compensation rate just equal to their marginal revenue product. The firm is assumed to operate in perfectly competitive markets. The manager is assumed to have perfect freedom with respect to varying his motivational effort among the fixed physical inputs. The coefficient β can be interpreted as the "technological efficiency share coefficient" of intermediate output. This coefficient need not be constant. Beta could possibly be a function of Z; i.e., $\beta(Z)$.

When confronted with a differential equation constraint, our manager cannot be myopic in his endeavor to maximize his utility index. If our manager-owner seeks to increase Y by redirecting more of his capacity effort to motivational functions--i.e., increase θ--current profit will increase. However, this action will decrease $\dot{Z}$ by decreasing $\dot{T}$. A lower $\dot{Z}$ implies that intermediate output and actual output will not grow as fast in the future, and that possibly profit may be reduced. Long-run profit is a function of T as well as θ; i.e.,

$$\pi = Y[\theta\bar{M}, Tf(\bar{X})] - \beta[Tf(\bar{X})]Tf(\bar{X})$$

Taking the time derivative of the long-run profit function, we get

$$\dot{\pi} = \frac{\partial Y}{\partial \theta}\dot{\theta} + \left(\frac{\partial Y}{\partial Z} - \beta'T - \beta\right)f(\bar{X})\dot{T}$$

Since T is a function of θ, we may want to consider the "instantaneous" effect of $\dot{\theta}$ on $\dot{T}$; i.e.,

$$\ddot{T} = -\frac{\partial g}{\partial \theta}\dot{\theta} + \frac{\partial g}{\partial E}\dot{T}f(\bar{X})$$

Although this type of instantaneous iteration could continue indefinitely, let

$$\dot{\pi} \approx \frac{\partial Y}{\partial \theta}\dot{\theta} - \left[\left(\frac{\partial Y}{\partial Z} - \beta' - \beta\right)f(\bar{X})\frac{\partial g}{\partial \theta}\right]\dot{\theta}$$

or

$$\frac{\dot{\pi}}{\dot{\theta}} = \frac{d\pi}{d\theta} \approx \frac{\partial Y}{\partial \theta} - \left(\frac{\partial Y}{\partial Z} - \beta' - \beta\right)f(\bar{X})\frac{\partial g}{\partial \theta}$$

Consequently, if the above approximation is negative, the profit level will decline in the future if our manager increases θ today.

The optimization problem facing our manager lends itself quite nicely to mathematical techniques dealing with optimal control problems; in particular, the approach utilizing the maximum principle of Pontryagin [3]. For all practical purposes, θ can be conceived as a "control" variable. The "state" variable is Z. Although Z is not observed, conceivably our manager has some cognizance of Z at any moment and can thus adjust his optimizing behavior accordingly.

Formally, the optimal control problem is expressed as:

$$\underset{\theta}{\text{Maximize:}} \quad V = \int_{t=0}^{\infty} U[Y(\theta,Z) - \beta(Z)Z]e^{-\rho t}\, dt$$

$$\text{Subject to:} \quad \dot{Z} = \psi(Y,Z),\ Z \geqq 0,\ Y \geqq 0$$

The manager maximizes the integral of discounted utility over an infinite horizon by finding an optimal time path of θ subject to the differential equation constraint and nonnegativity constraints. Forming the Hamiltonian function,

$$\kappa = U[Y(\theta,Z) - \beta(Z)Z]e^{-\rho t} + h\psi(Y,Z) \tag{9}$$

and making the following substitution, $q = he^{\rho}$, we obtain the modified Hamiltonian

$$H = e^{\rho t} = U[Y(\theta,Z) - \beta(Z)Z] + q\psi(Y,Z) \tag{10}$$

The necessary condition for maximization of the V-function is

$$\frac{\partial H}{\partial \theta} = U' \frac{\partial Y}{\partial \theta} + q\psi_Y \frac{\partial Y}{\partial \theta} = 0 \tag{11}$$

or

$$(U' + q\psi_Y) \frac{\partial Y}{\partial \theta} = 0 \tag{11a}$$

which implies

$$\frac{U'}{\psi_Y} = -q \tag{12}$$

Since $U' > 0$ and $\psi_Y < 0$, $q > 0$. The sufficient condition for maximum sum of utility is $\partial^2 H/\partial\theta^2 < 0$, which is met as expression (13) shows.

$$\frac{\partial^2 H}{\partial \theta^2} = U''\left(\frac{\partial Y}{\partial \theta}\right)^2 + U' \frac{\partial^2 Y}{\partial \theta^2} + q\psi_{YY}\left(\frac{\partial Y}{\partial \theta}\right)^2 + q\psi_Y \frac{\partial^2 Y}{\partial \theta^2} \tag{13}$$

We have $U'' < 0$, $\partial Y/\partial\theta > 0$, $\partial^2 Y/\partial\theta^2 < 0$, and $\psi_{YY} < 0$. From (12) we have $U' = -q\psi_Y$. Consequently, (13) can be reduced to

$$\frac{\partial^2 H}{\partial\theta^2} = U''\left(\frac{\partial Y}{\partial\theta}\right)^2 + q_{YY}\left(\frac{\partial Y}{\partial\theta}\right)^2 < 0 \tag{13a}$$

Applying the maximum principle algorithm, we obtain the following differential equations:

$$\dot{q} = U'(P''Z + P) - q\psi_Z + \rho q \tag{14}$$

and

$$\dot{Z} = \psi(Y,Z) \tag{15}$$

Differential equations (12), (14), and (15) enable us to solve for the optimal time paths of θ, Z, and Y.

We can now derive the differential equation for actual output Y. Taking the derivative of (1) with respect to time, we obtain

$$\dot{Y} = \frac{\partial Y}{\partial\theta}\dot{\theta} + \frac{\partial Y}{\partial Z}\dot{Z} \tag{16}$$

We know $\dot{Z}$, however we must derive the expression for $\dot{\theta}$. Take the derivative of $U' + q\psi_Y = 0$ with respect to time.

$$\left(U''\frac{\partial Y}{\partial\theta} + q\psi_{YY}\frac{\partial Y}{\partial\theta}\right)\dot{\theta} = -\psi_Y\dot{q} - \left[U''\frac{\partial Y}{\partial\theta} - \beta'Z - \beta\right) + q\left(\psi_{YY}\frac{\partial Y}{\partial Z} + \psi_{YZ}\right)\Big]\dot{Z}$$

Substituting equations (14) and (15) for $\dot{q}$ and $\dot{Z}$, respectively, and solving for $\dot{\theta}$, we have

$$\theta = \frac{-\psi_Y[\rho q + U'(\beta'Z + \beta) - q\psi_Z] - \psi[U''(\frac{\partial Y}{\partial Z} - \beta'Z - \beta) + q(\psi_{YY}\frac{\partial Y}{\partial Z} + \psi_{YZ})]}{U''\frac{\partial Y}{\partial\theta} + q_{YY}\frac{\partial Y}{\partial\theta}} \tag{17}$$

Consequently, substituting (15) and (17) for $\dot{Z}$ and $\dot{\theta}$, respectively, in (16), we can solve for the differential equation which describes the optimum time path of actual output.

$$\dot{Y} = \frac{U'\psi_Y(\rho - \psi_Z) + \psi_Y(U''\psi - U'\psi_Y)(\beta'Z + \beta) + U'\psi\psi_{YZ}}{U''\psi_Y - U'\psi_{YY}} \tag{18}$$

Differential equation (18) is a function of Y and Z. For convenience, we shall refer to (18) as the Ω-function; i.e., $\dot{Y} = \Omega(Y,Z)$. The Ω-function describes the behavior of the firm's actual output given an optimal θ policy.

We can now define what we shall refer to as the ϕ-function. The ϕ-function is differential equation (18) divided by differential equation (15); i.e.,

$$\frac{dY}{dZ} = \frac{\dot{Y}}{\dot{Z}} = \frac{\Omega(Y,Z)}{\psi(Y,Z)} = \phi(Y,Z)$$

or more specifically,

$$\frac{dY}{dZ} = \frac{U'\psi_Y(\rho - \psi_Z) + \psi_Y(U''\psi - U''\psi_Y)(\beta'Z + \beta) + U'\psi\psi_{YZ}}{\psi[U''\psi_Y - U'\psi_{YY}]} \tag{19}$$

Equation (19) is a parametric differential equation. If this ϕ-function is integrable, we can obtain a solution of the general form

$$Y^* = F(Z^*) \tag{20}$$

If $dY^*/dZ^* > 0$, we refer to (20) as the firm's optimal long-run production function. It should be noted that solutions of the form (20) represent a class of homothetic production functions.

§2.3 Analysis of Optimal Path

Since it is essential that $dY^*/dZ^* > 0$ for solutions of the form (20) to be considered long-run production functions, we should see if our general assumptions are sufficient to insure that any solution of the ϕ-function will be monotonically increasing. To do this, we assume that our firm has a stationary long-run optimum and that differential equation (19) describes the optimal path to this stationary state. Ideally, we would like to have the type of convergence path illustrated in Figure 1. In the case of most nontrivial dynamic programming problems, the convergence path is a saddle-point path.

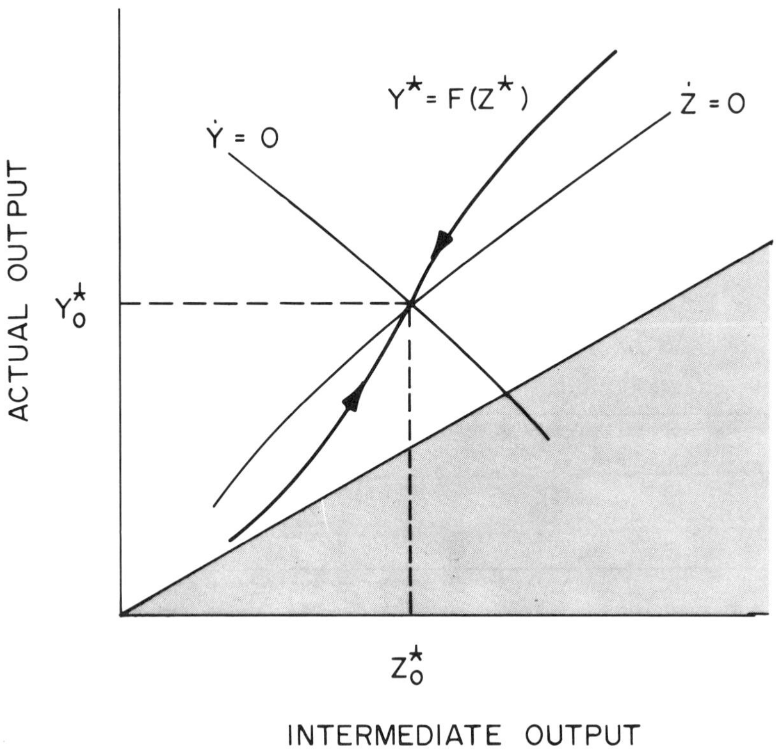

FIGURE 1. Desired convergence path.

Thus, we should determine what are the sufficient *qualitative* conditions for our desired path in terms of saddle-point solutions.

In general, we have a system of nonlinear differential equations composed of (18) and (15):

$$\dot{Y} = \Omega(Y,Z)$$
$$\dot{Z} = \psi(Y,Z)$$

A stationary equilibrium exists if

$$\dot{Y} = \Omega(Y_0^*,Z_0^*) = 0$$

and

$$\dot{Z} = \psi(Y_0^*,Z_0^*) = 0$$

Let $y = Y - Y_0^* \neq 0$ and $z = Z - Z_0^* \neq 0$ be small deviations from equilibrium. We wish to analyze the nonlinear system:

$$\dot{Y} = \Omega(Y,Z) = \Omega(Y_0^* + y,\ Z_0^* + z)$$
$$\dot{Z} = \psi(Y,Z) = \psi(Y_0^* + y,\ Z_0^* + z)$$

within some small neighborhood of equilibrium. Thus, we take linear approximations of the nonlinear differential equations. Let $\dot{y} = \Omega(Y_0^* + y,\ Z_0^* + z) - \Omega(Y_0^*,Z_0^*)$ and $\dot{z} = \psi(Y_0^* + y,\ Z_0^* + z) - \psi(Y_0^*,Z_0^*)$. From the abbreviated Taylor's series, we obtain the linearized system.

$$\dot{y} = \frac{\partial\Omega}{\partial Y}\,y + \frac{\partial\Omega}{\partial Z}\,z$$

$$\dot{z} = \frac{\partial\psi}{\partial Y}\,y + \frac{\partial\psi}{\partial Z}\,z$$

In vector notations we have

$$\begin{bmatrix} \dot{Y} \\ \dot{Z} \end{bmatrix} = \begin{bmatrix} \frac{\partial \Omega}{\partial Y} & \frac{\partial \Omega}{\partial Z} \\ \frac{\partial \psi}{\partial Y} & \frac{\partial \psi}{\partial Z} \end{bmatrix} \begin{bmatrix} Y \\ Z \end{bmatrix}$$

Let

$$A = \begin{bmatrix} a_{11} & a_{12} \\ a_{21} & a_{22} \end{bmatrix}$$

represent the matrix of first partials. The characteristic equation is defined by

$$\lambda^2 - (a_{11} + a_{22})\lambda + (a_{11}a_{22} - a_{12}a_{21}) = 0$$

If we let $u = a_{11} + a_{22}$ and $v = \det A = a_{11}a_{22} - a_{12}a_{21}$, we have

$$\lambda^2 - u\lambda + v = 0$$

The conditions for saddle-point convergence are (i) $u^2 - 4v > 0$, (ii) $v < 0$, and (iii) $u < 0$. Qualitatively, the following determinant satisfies the above conditions.

$$v = \begin{bmatrix} (-) & (-) \\ (-) & (+) \end{bmatrix}$$

Note that

$$u^2 - 4v = \left(\frac{\partial \Omega}{\partial Y} - \frac{\partial \psi}{\partial Z}\right)^2 + 4\,\frac{\partial \Omega}{\partial Z}\,\frac{\partial \psi}{\partial Y}$$

Thus $u^2 - 4v > 0$. We also have $u = \partial\Omega/\partial Y + \partial\psi/\partial Z < 0$. The sign restriction placed on the matrix of first partials *also insures that the saddle-point path is monotonically increasing.* This can be seen by examining the motion of the system.

Assuming $y \neq 0$ and $z \neq 0$, we set $\dot{y} = 0$ and $\dot{z} = 0$.

$$\frac{\partial \Omega}{\partial Y} dy + \frac{\partial \Omega}{\partial Z} dz = 0 \qquad \frac{\partial \psi}{\partial Y} dy + \frac{\partial \psi}{\partial Z} dz = 0$$

Therefore, for $\dot{y} = 0$,

$$\frac{dy}{dz} = - \frac{\partial \Omega}{\partial Z} \Big/ \frac{\partial \Omega}{\partial Y} < 0$$

and for $\dot{z} = 0$,

$$\frac{dy}{dz} = - \frac{\psi_Z}{\psi_Y} > 0$$

To the right of the $\dot{z} = 0$, $\dot{z} > 0$; and to the left of the curve, $\dot{z} < 0$. This implies that z increases to the right of curve and decreases to the left of the curve. To the right of the $\dot{y} = 0$ curve, $\dot{y} < 0$; and to the left, $\dot{y} > 0$. Therefore, y decreases to the right of the $\dot{y} = 0$ curve and increases to the left of the curve.

Combining all the vectors of motion of the system we have Figure 2. The stable trajectory probably looks like the solid curve. The broken curves are representative of unstable paths. Quantitatively, then, $\partial\Omega/\partial Y < 0$, $\partial\Omega/\partial Z < 0$, $\partial\psi/\partial Y < 0$, and $\partial\psi/\partial Z > 0$ are sufficient for $dY^*/dZ^* > 0$. We know the last two conditions are true. However, due to the complexity of the Ω-function, we cannot determine if the first two sign restrictions hold without some specific knowledge of the underlying functional forms. In the next two sections we will conduct an analysis utilizing some specific forms.

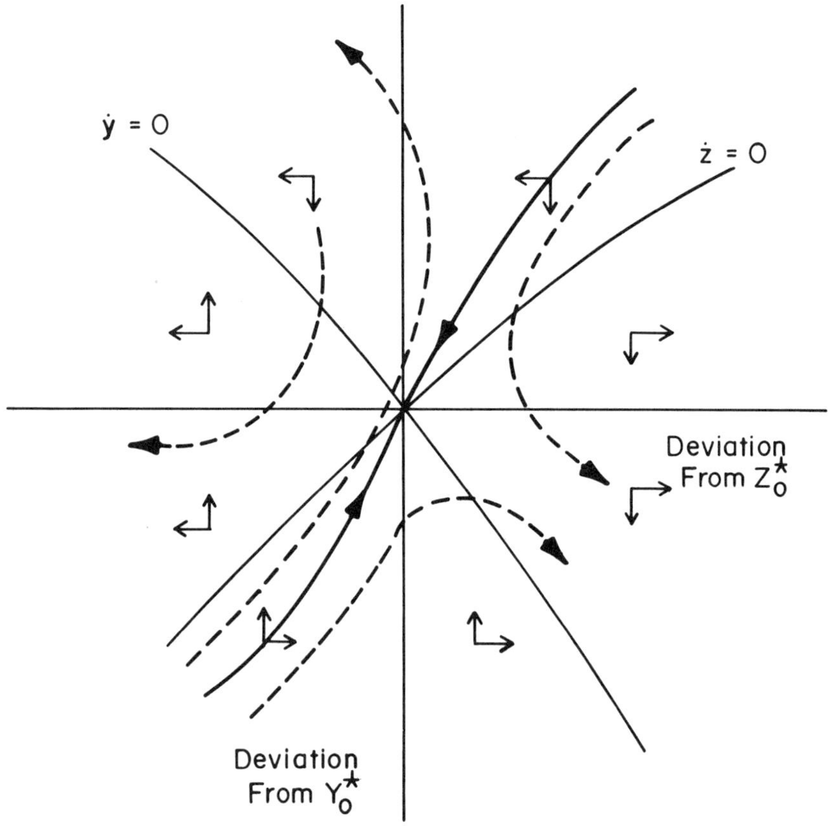

FIGURE 2. Stability of system.

§3. SPECIAL CASE I

§3.1 Underlying Relationships

Consider the following:

(i) $U(\pi) = a\pi + \frac{1}{2} b\pi^2 + c$ (quadratic utility function)

(ii) $\pi = Y(Z,\theta) - pZ$ (profit)

(iii) $\dot{Z} = \alpha Y + \beta Z + \gamma$ (linear dynamic constraint)

The quadratic utility function is increasingly being utilized in empirical research on preference behavior and is thus utilized for the first special case. In order for the marginal utility to be positive, a sufficiently large positive a parameter is required since diminishing marginal utility implies that b is negative. That is:

$$U'(\pi) = a + b\pi > 0 \qquad \text{(for } a > 0 \text{ and larger)}$$

and

$$U''(\pi) = b < 0$$

Expression (ii) defines profit. The technological efficiency share coefficient is assumed to be some constant p. From an economic perspective, we restrict profit to nonnegative values; i.e., $\pi \geqq 0$.

The linear dynamic constraint has a stringent economic intérpretation. To derive the linear constraint, assume that both the short-run production and the rate of change of intermediate output function are linear; i.e.,

$$\dot{Z} = rM_2 + sZ \qquad (21)$$

and

$$\dot{Y} = mM_1 + nZ \qquad (22)$$

where r, s, m, and n are all positive parameters. Since $M_1 + M_2 = \bar{M}$, we have $M_1 = \bar{M} - M_2$. Solving for M_1 in (22) we obtain $M_1 = Y - nZ/m$. Thus, by equality we have $\bar{M} - M_2 = Y - nZ/m$, which yields

$$M_2 = \frac{m\bar{M} + nZ - Y}{m} \qquad (23)$$

Substituting (23) into (21), we obtain

$$\text{(iiia)} \quad \dot{Z} = \left(-\frac{r}{M}\right)Y + \left(s + \frac{nr}{m}\right)Z + r\bar{M}$$

which is essentially our linear dynamic constraint where $\alpha = -\frac{r}{m}$, $\beta = s + \frac{nr}{m}$, and $\gamma = r\bar{M}$. The linear functions (21) and (22) imply that the marginal effectiveness of both managerial motivational effort and R&D effort are positive and constant; in other words, there are no diminishing returns to either types of managerial effort.

§3.2 Long-Run Production Function

We assume that our manager-owner seeks to maximize the integral of his discounted utility over an infinite horizon.

$$\text{Max } V = \int_{t=0}^{\infty} U(\pi)e^{-\rho t}\, dt$$

Subject to:

(i) $\dot{Z} = \psi(Y,Z)$

(ii) $Z \geqq 0$

(iii) $Y \geqq 0$

(iv) $0 \leqq \theta \leqq 1$

The control variable is θ, which denotes the proportion of our manager's capacity effort devoted to motivational activities. The relevant partial derivatives are:

$$U' = a + b\pi \geqq 0 \qquad \frac{\partial Y}{\partial \theta} > 0$$

$$U' = b < 0 \qquad \frac{\partial Y}{\partial Z} > 0$$

$$\psi_Y = \alpha > 0$$

$$\psi_Z = \beta > 0 \qquad \frac{\partial^2 Y}{\partial \theta^2} = 0$$

$$\psi_{YY} = \psi_{ZZ} = \psi_{YZ} = \psi_{ZY} = 0$$

The necessary condition for maximum sum of discounted utility is satisfied if

$$a + b\pi + q\alpha = 0 \tag{24}$$

for all t, where $q = he^{\rho t}$. The sufficient condition is readily satisfied since

$$\frac{\partial^2 \kappa}{\partial \theta^2} = b\left(\frac{\partial Y}{\partial \theta}\right)^2 < 0$$

Applying the maximum principle algorithm, we derive the differential equations which describe the optimal time paths of Y and Z. The system is composed of

$$\dot{Y} = \left[\frac{a(\rho - \beta - \alpha\rho)}{b} + p\gamma\right] + (\rho - \beta)Y + [\beta p - p(\rho - \beta - \alpha p)]Z \tag{25}$$

and

$$\dot{Z} = \gamma + \alpha Y + \beta Z \tag{26}$$

A useful property of this system of differential equations is that it is a linear system. For expository convenience let us utilize the following notations:

$$c_{11} = \rho - \beta \qquad c_{21} = \alpha$$

$$c_{12} = \beta p - p(\rho - \beta - \alpha p) \qquad c_{22} = \beta$$

$$c_{13} = \frac{a(\rho - \beta - \alpha p)}{b} + p\gamma \qquad c_{23} = \gamma$$

$$\dot{Y} = c_{11}Y + c_{12}Z + c_{13}$$

$$\dot{Z} = c_{21}Y + c_{22}Z + c_{23}$$

Our ϕ-function is described by

$$\frac{dY^*}{dZ^*} = \frac{c_{11}Y + c_{12}Z + c_{13}}{c_{21}Y + c_{22}Z + c_{23}} \tag{27}$$

In principle, there is a solution to differential equation (27), and this solution will yield the scale characteristics of the firm's long-run production function. In other words, we need some transversality condition in order to derive the exact solution to our optimality problem.

The transversality problem would be resolved if there exists a long-run *stationary* state which is determined by the parameters of the underlying relationships. If an optimum stationary state exists, then the optimum long-run values of Y and Z can be expressed as:

$$Y_0^* = \frac{\begin{vmatrix} -c_{13} & c_{12} \\ -c_{23} & c_{22} \end{vmatrix}}{\text{det. } A} \qquad Z_0^* = \frac{\begin{vmatrix} c_{11} & -c_{13} \\ c_{21} & -c_{23} \end{vmatrix}}{\text{det. } A}$$

where det. $A = c_{11} \cdot c_{22} - c_{21} \cdot c_{12}$. Solving for Y_0^* and Z_0^* explicitly in terms of the parameters of the system we obtain:

$$Y_0^* = \frac{-\frac{a\beta}{b} - \gamma p}{\beta + \alpha p} \qquad Z_0^* = -\frac{\gamma - \frac{\alpha a}{b}}{\beta + \alpha p}$$

If $Z_0^* > 0$, then $Y_0^* > 0$. This follows from our definition of profit; i.e., $\pi_0^* = Y_0^* - pZ_0^*$. Solving for π_0^*, we get $\pi_0^* = -a/b > 0$. Consequently, $Y_0^* = \pi_0^* + pZ_0^* > 0$. However, unless we have some knowledge of the magnitudes of the relevant parameters, we cannot determine if Z_0^* is positive or not.

For analytical purposes, we shall place restrictions (which conform to certain formal and economic criteria) on the underlying parameters and then examine the reasonableness of these restrictions. To begin with, we would like $Z_0^* > 0$. In addition, we require $F'(Z^*) > 0$ for economic reasons. And finally, we require that a stationary long-run equilibrium exists and is stable for at least one path. The behavior of our "shadow price" variable is described by

$$\dot{q} = a(\rho - \beta) + p(a + b\pi) \tag{28}$$

Since it is much simpler to deal with (28) than with (25), our analysis will be confined to the system composed of (26) and (28). This system describes the optimal behavior of the state variable Z and of the shadow price q.

Differential equation (28) is valid only if the necessary condition (24) holds. This requires that $q(t) = -\frac{a + b\pi}{\alpha}$ for every t. Substituting for q in (28) and consolidating terms, we get

$$\dot{q} = a\left[p - \frac{(\rho - \beta)}{\alpha}\right] + b\left[p - \frac{(\rho - \beta)}{\alpha}\right]\pi \tag{28a}$$

Whether $\dot{q}$ is a negative or positive function of π depends on the sign of the $(\alpha p - \rho + \beta)$ term. A stationary state solution requires that $\dot{q} = 0$. This implies that $\pi_0^* = -a/b$. It further implies that marginal utility is zero; i.e., $U(\pi_0^*) = a + b(-a/b) = 0$. Also, since q is a function of time (more specifically, $q = he^{\rho t}$), a stationary state requires that $q_0^* = 0$. This makes sense since $U'(\pi_0^*) = 0$. Using (24), we solve for $\pi = -\frac{\alpha q + a}{b}$. Substituting this expression into (28) and consolidating terms we obtain

$$\dot{q} = (\rho - \beta - \alpha p)q \tag{28b}$$

If $\rho - \beta - \alpha p \neq 0$, then q must equal zero in order that $\dot{q} = 0$.

Since $\pi = Y - pZ$, we can substitute $Y = \pi + pZ$ into differential equation (26).

$$\dot{Z} = \alpha\pi + (\alpha p + \beta)Z + \gamma$$

Letting $\pi = \pi_0^*$ and setting $\dot{Z} = 0$, we obtain

$$Z_0^* = \frac{\alpha a - \gamma b}{b(\alpha p + \beta)}$$

which confirms our earlier derivation of Z_0^*. Consequently, in terms of profit, shadow-price, and state variable, the stationary state is characterized by

$$\pi_0^* = -\frac{a}{b} > 0 \qquad q_0^* = 0 \qquad Z_0^* = \frac{\alpha a - \gamma b}{b(\alpha p + \beta)}$$

Let us now proceed to determine the restrictions required to yield a saddle-point convergence path to the stationary equilibrium. To do that, we analyze the differential equation system (26) and (28) in q and Z space. We therefore need

$$\dot{Z} = -\frac{\alpha^2}{b} q + (\alpha p + \beta)Z + \frac{\gamma b - \alpha a}{b}$$

In terms of small deviations from stationary equilibrium we have

$$\begin{bmatrix} \dot{q} \\ \dot{Z} \end{bmatrix} = \begin{bmatrix} \rho - \beta - \alpha & 0 \\ -\alpha^2/b & \alpha p + \beta \end{bmatrix} \begin{bmatrix} q^* - q_0^* \\ Z^* - Z_0^* \end{bmatrix}$$

We know $-\alpha^2/b > 0$. However, the signs of $(\rho - \beta - \alpha p)$ and $(\alpha p + \beta)$ are not ascertainable without knowledge of the magnitudes of the underlying parameters. Nevertheless, for stability purposes, the sign of these two coefficients cannot be arbitrary. Given the conviction that the firm's long-run behavior is stable in some sense, we can proceed to determine the sign of these two coefficients.

First, let us examine $\rho - \beta - \alpha p$. Since $\alpha p < 0$, $\rho - \beta$ would have to be negative in order that $\rho - \beta - \alpha p < 0$. On the other hand, if $\rho - \beta > 0$, then $\rho - \beta - \alpha p > 0$. The situation where $\rho - \beta - \alpha p > 0$ is undesirable from the stability standpoint, for it implies that as long as q is positive, q will tend to increase rather than decrease. Since we require $q_0^* = 0$, this situation is destabilizing. Consequently, we assume that $\rho - \beta - \alpha p < 0$.

Saddle-point stability requires that both characteristic roots of the system be real and that they be of opposite signs. The traced is positive since $p > 0$. The determinant of the matrix of coefficients is

$$\text{det. } A = (\rho - \beta - \alpha p)(\alpha p + \beta) = \rho(\alpha p + \beta) - (\alpha p + \beta)^2$$

If $\alpha p + \beta > 0$, it immediately follows that the discriminant is positive and that the characteristic roots are real and of the opposite sign; i.e.,

$$\lambda_1 = -(\alpha p + \beta) < 0$$

and

$$\lambda_2 = -(\rho - \beta - \alpha p) > 0$$

Thus, the saddle-point requirements are satisfied. Note that if $\alpha p + \beta < 0$, the stationary equilibrium will be completely unstable.

We now examine if $F'(Z^*) > 0$ is satisfied. To do that, we must return to our differential equation system composed of equations (25) and (26). Saddle-point convergence requires (i) $\rho - \beta - \alpha p < 0$ and (ii) $\alpha p + \beta > 0$. Requirement (i) implies that $\rho - \beta < 0$. Consequently, $c_{11} < 0$. We have $c_{12} > 0$ from requirement (ii). Since $c_{21} < 0$ and $c_{22} > 0$, we have the following qualitative matrix:

$$\begin{bmatrix} (-) & (+) \\ (-) & (+) \end{bmatrix}$$

Since this sign pattern does not conform to the sufficient condition presented in Section 2, we have to determine if the convergence path traces out a function where Y is a monotone increasing function of Z. If the saddle-point convergence requirements are satisfied for the q and Z space, they are also satisfied for the Y and Z space. In terms of the motion of the system, there are two possible cases in the Y and Z space. These are represented in Figure 3. In both cases, if a convergence path exists, then it also satisfies the requirement that $F'(Z^*) > 0$.

The restrictions placed on the parameters for saddle-point convergence are not sufficient to satisfy the requirement that $Z_0^* > 0$. For $Y_0^* > 0$ and $Z_0^* > 0$, we require the additional restriction (iii) $\gamma - \frac{\alpha a}{b} < 0$. This follows from examination of $Z_0^* = \frac{\alpha a - \gamma b}{b(\alpha p + \beta)}$. *Thus, a saddle-point stationary equilibrium will exist and satisfy the requirement that* $F'(Z^*) > 0$, *if the following conditions prevail:*

(i) $\rho - \beta - \alpha p < 0$

(ii) $\alpha p + \beta > 0$

(iii) $\gamma - \frac{\alpha a}{b} < 0$

Except for (i), the above conditions appear to be reasonable in terms of the kinds of relationships we are dealing with.

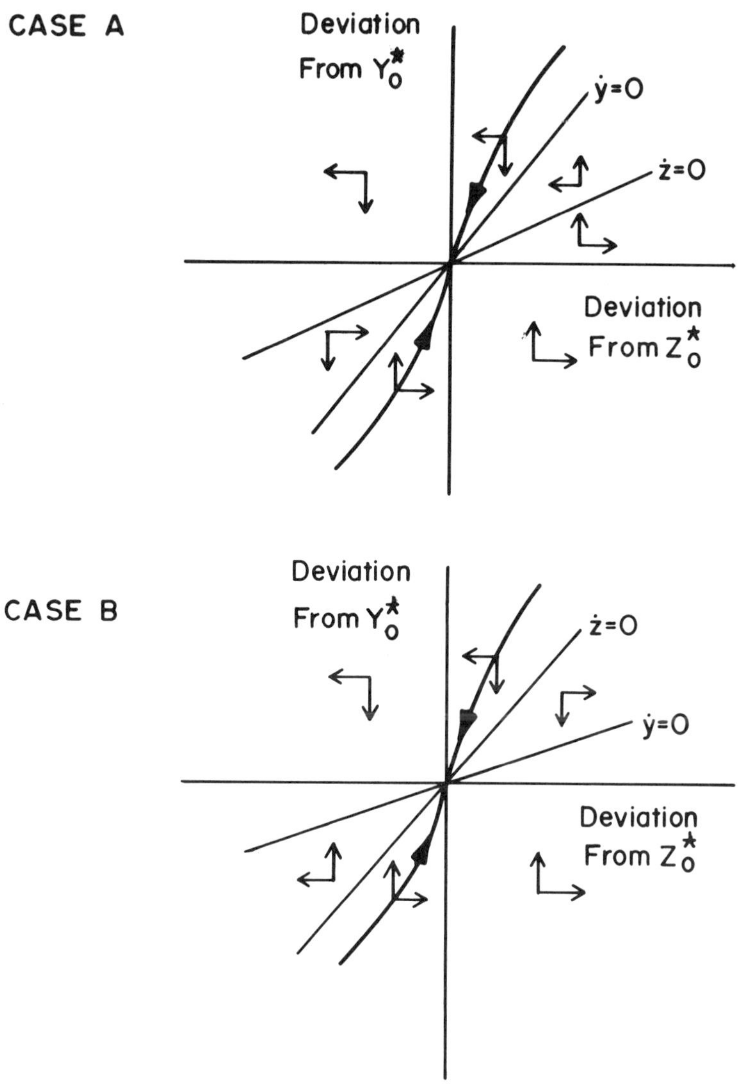

FIGURE 3. Convergence for Special Case I.

Since it was already pointed out that we need a significantly large positive a parameter in order for marginal utility to be positive over a substantial range of profit, condition (iii) would appear to be plausible. Condition (ii) appears to be reasonable, more so if p, the technological efficiency share coefficient, is closer to zero than to one. Condition (i) implies that $\rho - \beta < 0$ and that β is substantially larger than ρ, the subjective discount rate. Of all three conditions, this particular condition appears to be the one which is least likely to prevail. As was pointed out earlier, $\beta = s + \frac{nr}{m}$, where $\partial\dot{Z}/\partial Z = s$, $\partial\dot{Z}/\partial M_2 = r$, $\partial Y/\partial M_1 = m$, and $\partial Y/\partial Z = n$. Thus, β essentially depends on the marginal effectiveness of the level of intermediate output in stimulating the rate of technological advance. One would expect s to be very small. Consequently, if the manager-owner has any time preference, condition (i) will most likely not be true.

§3.3 Comparative Stationary State

In this section we evaluate the effects of parameter changes on Y_0^*, Z_0^*, and π_0^*. We have seven parameters: a, b, α, β, σ, ρ, and p. All except the discount rate ρ enter into the determination of the stationary state values. The effect of changes in these parameters on the stationary state values are summarized in Table 1. It is interesting to note that the stationary state profit level is subject only to changes in the utility function parameters. An increase in a or a decrease in b yields a higher optimum long-run profit level. This is not unexpected, for $U'(\pi_0^*) = 0$ essentially determines

TABLE 1. Summary of Comparative Stationary States Results

Parameter	Parametric effect on stationary state value Y_0^*	Z_0^*	π_0^*
a	$\frac{-\beta}{b(\beta + \alpha p)} > 0$	$\frac{\alpha}{b(\beta + \alpha p)} > 0$	$-\frac{1}{b} > 0$
b	$\frac{-\gamma pb(\beta + \alpha p) + (a\beta + \gamma pb)(\beta + \alpha p)}{[b(\beta + \alpha p)]^2} > 0$	$\frac{-\gamma b(\beta + \alpha p) - (\alpha a \quad b\gamma)(\beta + \alpha p)}{[b(\beta + \alpha p)]^2} > 0$	$\frac{a}{b^2} > 0$
α	$\frac{(a\beta + \gamma pb)bp}{[b(\beta + \alpha p)]^2} < 0$	$\frac{b(a\beta + pb\gamma)}{[b(\beta + \alpha p)]^2} < 0$	None
β	$\frac{bp(\gamma b - a\alpha)}{[b(\beta + \alpha p)]^2} < 0$	$\frac{-(\alpha a - b\gamma)b}{[b(\beta + \alpha p)]^2} < 0$	None
γ	$\frac{-pb}{b(\beta + \alpha p)} < 0$	$\frac{-b}{b(\beta + \alpha p)} < 0$	None
p	$\frac{b(a\alpha - \gamma b)}{[b(\beta + \alpha p)]^2} > 0$	$\frac{-(\alpha a - b\gamma)b}{[b(\beta + \alpha p)]^2} > 0$	None

the long-run stationary state position with regard to the firm's profit level. An increase in $\underline{p}$ results in a proportional increase in Y_0^* and Z_0^*. A decrease in α, or an increase in either β or γ, results in a proportional decline in Y_0^* and Z_0^*.

§3.4 Comparative Dynamics

In addition to evaluating the effects of parameter changes on the stationary state solution, we can also examine how parameter changes alter the shape of the optimal path. As a characteristic of the optimal path, we choose its slope, which is defined by our ϕ-function. The ϕ-function has the following implicit solution:

$$\int(c_{21}Y + c_{22}Z + c_{23})dY - \int(c_{11}Y + c_{12}Z + c_{13})dZ + k = 0$$

$G(Y^*,Z^*,k) = 0$, where k reflects the constant of integration terms. In the long-run, $G(Y^*,Z^*,c) \rightarrow G(Y_0^*,Z_0^*)$. Since the slope of the optimal path is not independent of Y^* and Z^*, we select $\bar{Y}^*$ and $\bar{Z}^*$ such that $\bar{Y}^* < Y_0^*$ and $\bar{Z}^* < Z_0^*$, and evaluate the effects of parameter changes on the ϕ-function at these selected values. Thus, the slope of our optimal path at $\bar{Y}^*$ and $\bar{Z}^*$ is expressed as

$$\phi(\bar{Y}^*,\bar{Z}^*) = \frac{(\rho - \beta)\bar{Y}^* + [\beta p - p(\rho - \beta - \alpha p)]\bar{Z}^* + \dfrac{a(\rho - \beta - \alpha p)}{b} + p\gamma}{\alpha\bar{Y}^* + \beta\bar{Z}^* + \gamma}$$

Both the numerator (N) and denominator (D) of $\phi(\bar{Y}^*,\bar{Z}^*)$ are positive since $\bar{Y}^* < Y_0^*$ and $\bar{Z}^* < Z_0^*$.

The effect of changes in either α or $\underline{p}$ on the shape of the optimal path are indeterminate.

$$\frac{\partial\phi}{\partial\alpha} = \frac{(p^2\bar{Z} - \frac{ap}{b})(D) - (N)\bar{Y}*}{(D)^2} \qquad \text{(sign?)}$$

$$\frac{\partial\phi}{\partial p} = \frac{[\beta + \alpha(\rho - \beta - \alpha p)]\bar{Z}* + (\gamma - \frac{a\alpha}{b})}{D} \qquad \text{(sign?)}$$

An increase in a unambiguously increases the slope of the optimal path at $\bar{Y}*$ and $\bar{Z}*$, while a decrease in b unambiguously decreases the slope, i.e.,

$$\frac{\partial\phi}{\partial a} = \frac{\frac{\rho - \beta - \alpha p}{B}}{D} > 0 \qquad \frac{\partial\phi}{\partial b} = \frac{\frac{a(\rho - \beta - \alpha p)}{b^2}}{D} < 0$$

An increase in β will decrease the slope if $\pi_0^* - \bar{\pi}* + p\bar{Z}* < 0$.

$$\frac{\partial\phi}{\partial\beta} = \frac{(-\bar{Y}* + 2p\bar{Z}* - \frac{a}{b})(D) - (N)\bar{Z}*}{(D)^2}$$

Since $-a/b = \pi_0^*$, we have $\pi_0^* - (\bar{Y}* - p\bar{Z}*) + p\bar{Z}*$. Let $\bar{\pi}* = \bar{Y}* - p\bar{Z}*$. Consequently,

$$\frac{\partial\phi}{\partial\beta} = \frac{(\pi_0^* - \bar{\pi}* + p\bar{Z}*)(D) - (N)\bar{Z}*}{(D)^2} < 0, \text{ if } \pi_0^* - \bar{\pi}* + p\bar{Z}* > 0$$

An increase in γ will increase the slope of the optimal path if $\pi_0^* + \bar{\pi}* - 2\bar{Y}* < 0$ and will decrease the slope if $\pi_0^* + \bar{\pi}* - 2\bar{Y}* > 0$.

$$\frac{\partial\phi}{\partial\beta} = \frac{p(D) - (N)}{(D)^2}$$

$$= \frac{(-\bar{Y}* - p\bar{Z}* - \frac{a}{b})(\rho - \beta - \alpha p)}{(D)^2}$$

$$= \frac{(\pi_0^* + \bar{\pi}* - 2\bar{Y}*)(\rho - \beta - \alpha p)}{(D)^2} \begin{matrix} > 0, \text{ if } \pi_0^* + \bar{\pi}* - 2\bar{Y}* < 0 \\ < 0, \text{ if } \pi_0^* + \bar{\pi}* - 2\bar{Y}* > 0 \end{matrix}$$

Finally, an increase in the discount rate ρ will increase the slope if $\bar{\pi}^* - \pi_0^* > 0$ and will decrease the slope if $\bar{\pi}^* - \pi_0^* < 0$.

$$\frac{\partial\phi}{\partial p} = \frac{\bar{Y}^* - p\bar{Z}^* + \frac{a}{b}}{D} \frac{\bar{\pi}^* - \pi_0^*}{D} \begin{cases} > 0, \text{ if } \bar{\pi}^* - \pi_0^* > 0 \\ < 0, \text{ if } \bar{\pi}^* - \pi_0^* < 0 \end{cases}$$

§4. SPECIAL CASE II

§4.1 Underlying Relationships

Except for the form of the utility function, the underlying relationships of Special Case II are the same as those of Special Case I. In Special Case II we assume an isoelastic utility function. Thus, in place of (i) we substitute

$$u = \pi^a \qquad (0 < a < 1) \tag{29}$$

A distinguishing feature of the isoelastic utility function is that it has no saturation point; i.e., $u'(\pi) \neq 0$ for $\pi > 0$. This contrasts with the case of the quadratic utility function which does have a saturation point. It will be shown later that these two contrasting utility functions imply contrasting shadow price behavior

As in our previous special case, we assume that our manager seeks to maximize the integral of his discounted utility over an infinite horizon. By applying the maximum principle a logarithm, we derive the system of differential equations which describe the optimal dynamic behavior of Y and Z, respectively:

$$\dot{Y} = [p\alpha + \frac{(\rho - \beta) - \alpha p}{a - 1}]Y + [p\beta - \frac{p(\rho - \beta - \alpha p)}{a - 1}]Z + p\gamma \tag{30}$$

and

$$\dot{Z} = \alpha Y + \beta Z + \gamma \tag{31}$$

As in Special Case I, we have a system of linear differential equations. Unlike the first case, however, this system does not yield a stationary state solution with positive optimal profit.

Setting (30) and (31) equal to zero and solving for Y_0^* and Z_0^* simultaneously, we get:

$$Y_0^* = - \frac{\gamma p}{\alpha p + \beta} \qquad \text{and} \qquad Z_0^* = - \frac{\gamma}{\alpha p + \beta}$$

By our definition of profit we have

$$\pi_0^* = Y_0^* - pZ_0^* = - \frac{\gamma p}{\alpha p + \beta} + \frac{\gamma p}{\alpha p + \beta} = 0$$

Consequently, if an optimal stationary state exists, then it must be at zero profit.

To determine if a stationary state is feasible, let us examine the behavior of q. From the necessary condition for maximization of the integral of discounted utility we have

$$q = - \frac{u'(\pi)}{\psi_Y} = - \frac{a\pi^{a-1}}{\alpha}$$

where $q = he^{\rho t}$. Optimal dynamic behavior of q is characterized by[4]

$$\dot{q} = q(\rho - \beta) + pa\pi^{a-1} \tag{32}$$

---

[4]This follows from: $\dot{q} = \rho q - (u' + q\psi_Y)\frac{\partial Y}{\partial Z} + u'p - q\psi_Z$.

or

$$\dot{q} = a[p - \frac{\rho - \beta}{\alpha}]\pi^{a-1} \tag{32a}$$

For convenience, let $C = a[p - \frac{\rho - \beta}{\alpha}]$. Therefore,

$$\dot{q} = \frac{C}{\pi^{1-a}} \tag{32b}$$

By examination, it is obvious that unless $C = 0$, $\dot{q}$ can never be zero. Consequently, Special Case II does not have a feasible stationary state.

§4.2 Comparison of Shadow Price Behavior

The two special cases analyzed in this section and the previous one imply quite different shadow price behavior. In the first case, $\dot{q}$ *increases* as π increases, while in the second case $\dot{q}$ increases as π *decreases*. In Special Case I, it is feasible to have $\dot{q} = 0$ for $\pi > 0$. However, in Special Case II, $\dot{q}$ approaches 0 as π increases, but is never zero.

Formally, the variable q represents the undiscounted Lagrangian multiplier; i.e., $q = he^{\rho t}$. Thus, q is not independent of time unless $h = 0$. The necessary condition for optimal behavior requires

$$q = -\frac{u'(\pi)}{\psi_Y} \qquad \text{or} \qquad h = -\frac{u'(\pi)e^{\rho t}}{\psi_Y}$$

In other words, optimality requires that the manager increase his motivational effort up to the point where the marginal change in the time rate of change in intermediate output just equals his discounted marginal utility per unit of h. Thus, we can interpret h to be the discounted subjective price of the manager's motivational effort and q to be the undiscounted subjective price.

In terms of the above interpretation of q, the dynamic shadow price behavior of Special Case I appears to be meaningful. As profit is approaching the stationary state value, the undiscounted subjective price is falling at a slower and slower rate; i.e., $\dot{q} < 0$ ($\pi^* < \pi_0^*$) and $d\dot{q}/d\pi > 0$. On the other hand, in Special Case II the subjective price is *rising* when profit is increasing; which is intuitively unrealistic.

REFERENCES

1. Clemhout, S., The Class of Homothetic Isoquant Production Functions, Review of Economic Studies, 35, 91-104 (Jan. 1968).
2. Meyer, P. A., An Aggregate Homothetic Production Function, Southern Economic Review, 36, 229-238 (Jan. 1970).
3. Pontryagin, L. S., Boltyanskii, V. G., Gamkrelidze, R. V., and Mischenko, E. F., The Mathematical Theory of Optimal Processes. English translation by K. N. Trirogoff, Interscience, New York, 1962.
4. Sato, R., Koizumi, T., and Wolkowitz, B., Market Behavior and the Types of Production Functions, European Economic Review, 6, 331-342 (Oct. 1975).
5. Shepard, R. W., Cost and Production Functions, Princeton University Press, Princeton, 1953.
6. Zellner, A., and Revankar, N. S., Generalized Production Functions, Review of Economic Studies, 36, 241-250 (April 1969).

DYNAMIC MODELS OF FISHING:
A HEURISTIC APPROACH

Leonard J. Mirman

Department of Economics
University of Illinois
Urbana-Champaign, Illinois

§1. INTRODUCTION

In this paper models of the interaction of biological and economic aspects of fishing or any renewable (or even exhaustible) resource will be studied. All of the models studied have one common attribute, namely that the underlying population, fish for ease in exposition, is dynamic. The next two sections of this paper deal with the problem in which countries are interested in the welfare of their citizens. In particular, we assume in these sections that the countries maximize the sum of discounted utilities subject to a constraint imposed by the nature of the biological growth function. The assumptions which are made in the next two sections are as follows. Let c_t be the consumption of fish in period t, then $u(c_t)$ is the utility derived from the consumption c_t. It is assumed that $u' > 0$, $u'' < 0$, and that $u'(0) = +\infty$. Let x_t be the amount of fish in period t; then, if no fishing is done

$$x_{t+1} = f(x_t)$$

where $f(0) = 0$, $f' > 0$, $f'' < 0$, and $f'(0) = \infty$. The function f is the biological growth function of fish. The last assumption is somewhat unrealistic, but in the first two sections of the paper it guarantees that a positive amount of fish can be sustained, and will be sustained under an optimal fishing policy. This function is illustrated in Figure 1.

The difference between Section 2 and 3 is that the assumption made in the first section is of a central planner maximizing the welfare in his own country (or a combination of the welfare in two countries). Optimal policies are derived for an example. The interesting question then arises; suppose that there is more than one country--in particular, two countries--how will the competition between them affect the size of the catch and the eventual size of the fish population? This

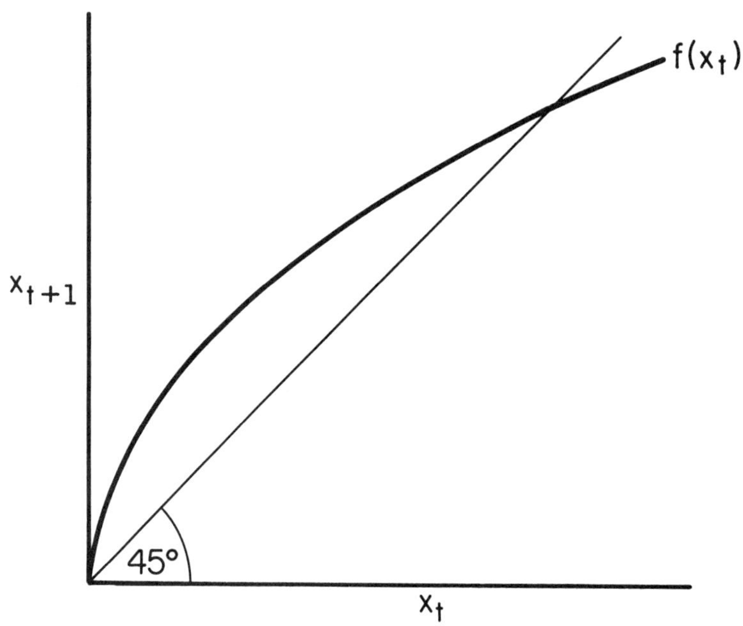

FIGURE 1

question is treated in the third section. It is shown using a dynamic Cournot-Nash equilibrium that the duopolists consume more fish, for each level of fish, than does a monopolist.

Sections 4 and 5 of the paper deal with the same type of questions considered in the first two sections except that it is assumed that profit maximization is the objective rather than utility maximization. It is assumed that the demand curve is given by D(q), where q is the quantity of fish sold. Further, it is assumed that there are no costs. These conditions are assumed to remain constant over time. Finally in these sections the dynamics of fish is assumed to be linear, i.e.,

$$x_{t+1} = rx_t$$

where $r > 1$. The dynamics are depicted in Figure 2.

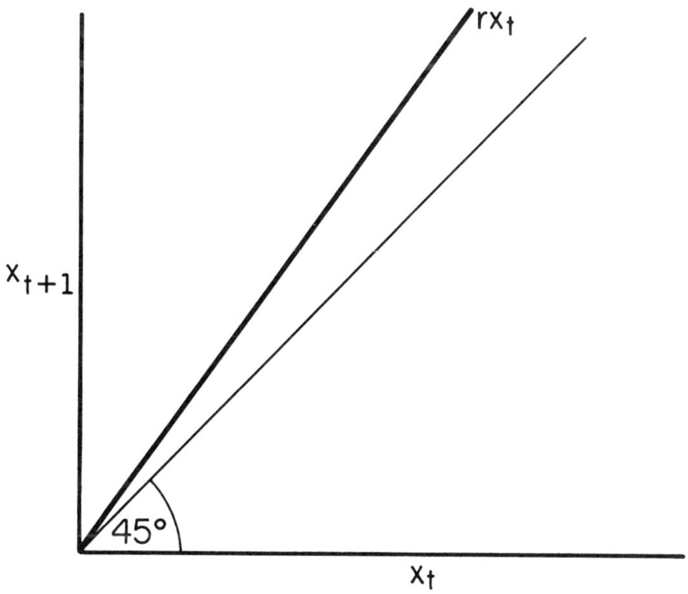

FIGURE 2

Section 4 on profit maximization is of a monopolist which maximizes profits while Section 5 studies the duopoly problem. It is shown in this section that less fish will be caught by the monopolist than by the two duopolists. For the profit maximization case the steady state equilibrium is a market equilibrium which is exactly the same as in the static model since the optimal static market supply may be met while the population of fish is sustained. In the final section of this paper the shortcomings of Sections 3 and 5 are investigated. In particular, an example is used to show that the usual dynamic programming results are not always available in the Cournot-Nash equilibrium case.

§2. UTILITY MAXIMIZATION UNDER COOPERATION (OR MONOPOLY)

In this section each country is assumed to maximize the sum of discounted utilities of consumption. First, the case of a single country will be discussed; however, this could be thought of as maximizing a convex combination of both countries' utilities. Both countries are assumed to have identical time preference or discount factors δ, $0 < \delta < 1$, and identical utility functions. Let

$$V(s) = \sup_{\{c_t\}} \sum_{t=0}^{\infty} \delta^t u(c_t)$$

subject to

$$x_0 + c_0 = s$$

and

$$x_t + c_t = f(x_{t-1}) \qquad t = 1, 2, 3, \ldots$$

Then using the dynamic programming formulation, V(s) may be written in the form,

$$V(s) = \max_{c}\{u(c) + \delta V(f(s - c))\} \tag{1}$$

For this problem it can be shown that V is an increasing, concave, differentiable function, hence the maximum is proper in equation (1). For properties of the function V in this case, see Mirman and Zilcha [5]. Moreover the first order conditions from the maximization problem (1) has a unique solution. This condition is

$$u'(c) = \delta V'(f(s - c))f'(s - c) \tag{2}$$

From (1) letting the optimal policy c be a function of s, i.e., suppose c = g(s), then

$$V(s) = u(g(s)) + \delta V(f(s - g(s))) \tag{3}$$

Differentiating (3) with respect to s yields

$$V'(s) = u'(g(s))g'(s) + \delta V'(f(s - g(s))) \cdot f'(s - g(s))(1 - g'(s)) \tag{4}$$

Using equation (2),

$$V'(s) = \delta V'(f(s - g(s)))f'(s - g(s)) \tag{5}$$

This yields the important envelope result, obtained by substituting (5) into (2),

$$u'(g(s)) = V'(s) \tag{6}$$

The interpretation of this equation is that the value of another fish is merely the marginal utility of consumption.

The envelope result can be used to derive the discrete time Euler equations. Namely,

$$u'(g(s)) = \delta u'(g(f(s - g(s))))f'(s - g(s)) \tag{7}$$

or, in general,

$$u'(g(f(x_t))) = \delta u'(g(f(x_{t+1})))f'(x_{t+1}) \tag{8}$$

Using (8), the dynamics of the process x_t as well as the steady state condition can be studied.

In particular the steady state properties of this system may be compared to its Cournot-Nash analog which will be studied in the next section.

Definition The process x_t is said to be in the *steady state* if

$$x_t = x_{t+1} \quad \text{for all } t \tag{9}$$

In the steady state the number of fish, as well as the fish consumption, in each period is identical, hence

$$u'(g(f(x_t))) = u'(g(f(x_{t+1}))) \quad \text{for all } t$$

Thus for a steady state solution it follows from equation (8) that

$$1 = \delta f'(\bar{x})$$

Since the dynamics of the Cournot, or duopoly, case cannot be studied in general, it is most convenient at this point to study the behavior of this optimal process in terms of an example, since this is the best way to compare the conclusions of each of the models. Hence, let

$$u(c) = \log c$$

and

$$f(x) = x^{\alpha} \qquad 0 < \alpha < 1$$

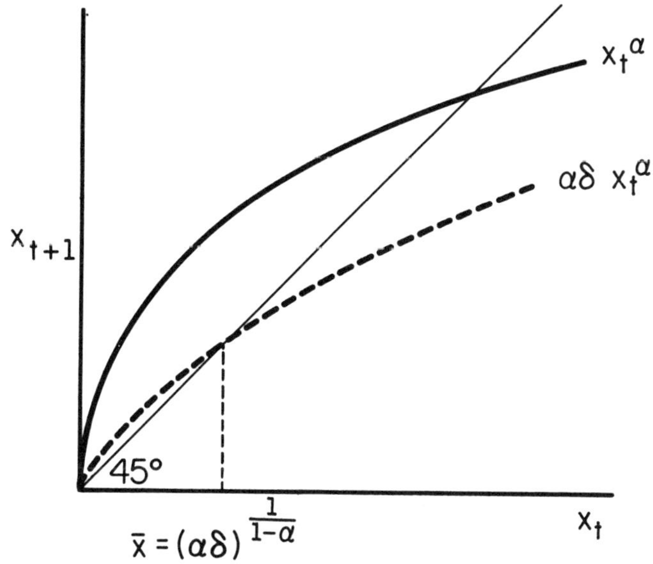

FIGURE 3

Equation (8) may be used to derive the dynamics for this example, in particular,

$$\frac{1}{g(s)} = \frac{\alpha\delta(s - g(s))^{\alpha-1}}{g[(s - g(s))^{\alpha}]}$$

As a solution, try $g(s) = \lambda s$, then

$$\frac{1}{\lambda s} = \frac{\alpha\delta(1 - \lambda)^{\alpha-1}s^{\alpha-1}}{\lambda(1 - \lambda)^{\alpha}s^{\alpha}}$$

Thus

$$1 = \frac{\alpha\delta}{1 - \lambda} \quad \text{or} \quad 1 - \lambda = \alpha\delta, \quad \lambda_M = 1 - \alpha\delta$$

From the optimal policy function $g(s) = (1 - \alpha\delta)s$, the dynamics of the system may be found. In fact

$$x_{t+1} = f(x_t) - g(f(x_t)) = x_t^{\alpha} - (1 - \alpha\delta)x_t^{\alpha}$$
$$= \alpha\delta x_t^{\alpha}$$

Hence, for a steady state,

$$\bar{x} = \alpha\delta\bar{x}^{\alpha} \qquad \bar{x}^{1-\alpha} = \alpha\delta \qquad \bar{x} = \alpha\delta^{\frac{1}{1-\alpha}}$$

The system is depicted in Figure 3.

§3. UTILITY MAXIMIZATION UNDER DUOPOLY

In this section each country is again assumed to maximize the sum of discounted utilities of consumption, and each country is required to take account of the actions of the other country. It will be assumed that both countries are identical in the sense that they both have the same utility function as well as the same discount factor. The value function is given by

$$V(s) = \sup_{c_t} \sum_{t=0}^{\infty} \delta^t u(c_t)$$

subject to

$$x_t + c_t + h(f(x_{t-1})) = f(x_{t-1}) \qquad t = 1, \ldots$$

and

$$y_0 + c_0 + h(s) = s$$

Here h is the policy function of the other country. Using a dynamic programming formulation, we obtain

$$V(s) = \max_{c}\{u(c) + \delta V(f(s - c - h(s)))\} \tag{10}$$

At this point it would be nice to say that all properties which are valid in the cooperative case are also valid here. However, this is not in general true. In particular it is not possible to say that V is concave, as will be shown by an example below, nor does it necessarily follow that a maximum, as written in (10), is achieved. Finally, the properties of h, which are endogenous and which usually hold in the optimal case, may not hold here. For example, the optimal policy may not be continuous or increasing. Again this will be shown in the example below. (Although in this example the right hand and left hand derivatives are, as one would expect, between zero and one.) We shall proceed as if all properties which are needed hold. In the example presented in this section they will in fact hold. However, the counterexample will show these properties should not always be expected to hold.

From (10) the first order condition is

$$u'(c) = \delta V'(f(s - c - h(s)))f'(s - c - h(s)) \tag{11}$$

Letting c = g(s) be the optimal policy given h(s), (10) may be rewritten as

$$V(s) = u(g(s)) + \delta V(f(s - g(s) - h(s))) \tag{12}$$

Differentiating with respect to s yields

$$V'(s) = u'(g(s))g'(s) + \delta V'(f(s - g(s) - h(s))) \cdot f'(s - g(s) - h(s))(1 - g'(s) - h'(s)) \tag{13}$$

Now the first order condition (11) may be used to show

$$V'(s) = \delta V'(f(s - g(s) - h(s))) \cdot f'(s - g(s) - h(s))(1 - h'(s)) \tag{14}$$

This equation should be compared to (5) in which the term

$1 - h'(s)$ does not appear. Again an "envelope" result can be achieved. However, as is clear from (14), in this case the term $1 - h'(s)$ must be incorporated, namely,

$$V'(s) = u'(g(s))(1 - h'(s)) \tag{15}$$

The interpretation of this equation is that the value of another fish is the marginal utility of consumption multiplied by the portion of the fish left after the other countries' consumption is considered. As in the cooperative case, one would expect that $1 - h'(s)$ is positive and less than one. It is at precisely this point that the difficulty with the general analysis is manifested. In particular, using the envelope theorem in the cooperative case, it follows that

$$V''(s) = u''g'(s)$$

(V'' does not in general exist.) Using other arguments, it can be shown that $g' > 0$, hence $V'' < 0$. However, in the Cournot-Nash case differentiation of (15) yields

$$V''(s) = u''g'(1 - h') - u'h''$$

The difficulty is that h'' can have any sign and magnitude. Hence V may not be concave. Again this possibility will be shown in the counterexample below.

The "envelope" relationship (15) may be used to find the discrete Euler equation

$$u'(g(s)) = \delta u'(g(f(s - g(s) - h(s))))f'(s - g(s) - h(s)) \cdot (1 - h'(f(s - g(s) - h(s))) \tag{16}$$

or in general

$$u'(g(f(x_t))) = \delta u'(g(f(x_{t+1})))f'(x_{t+1})(1 - h'(f(x_{t+1}))) \tag{17}$$

from which the motion of the Cournot-Nash equilibrium may be studied. This equation is difficult to study in general since the term $1 - h'(f(s - g(s) - h(s)))$ complicates matters. However, in the context of an example several interesting insights may be obtained, since for the example h' is a constant. Before discussing the example the steady state condition which does yield some further insights will be studied.

In the steady state, output and, therefore, consumption must be the same in each period. Hence, for a steady state,

$$1 = \delta f'(\bar{\bar{x}})(1 - h'(f(\bar{\bar{x}})))$$

Under the properties usually associated with optimal consumption functions, i.e., $0 < 1 - h' < 1$,

$$\delta f'(\bar{\bar{x}}) > 1 \qquad \text{and} \qquad \bar{\bar{x}} < \bar{x}$$

since f is concave. This comparison is depicted in Figure 4.

Consider now the example studied in the previous section. Again, let $u(c) = \log c$, and $f(x) = x^\alpha$, $0 < \alpha < 1$. For this example, equation (17) becomes

$$\frac{1}{g(s)} = \frac{\alpha\delta(s - g(s) - h(s))^{\alpha-1}(1 - h'[(s - g(s) - h(s))^\alpha])}{g[(s - g(s) - h(s))^\alpha]} \tag{18}$$

Now suppose $g(s) = h(s) = \lambda s$, i.e., assume that in equilibrium both countries have the same linear Cournot-Nash policy. Under this assumption, (18) becomes

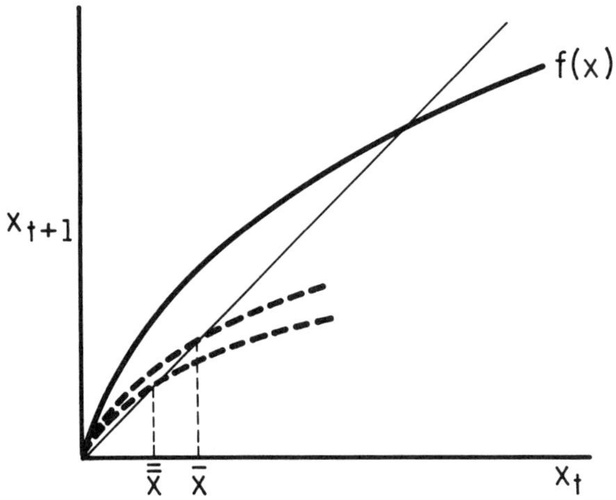

FIGURE 4

$$\frac{1}{\lambda s} = \frac{\alpha\delta(1 - 2\lambda)^{\alpha-1}s^{\alpha-1}(1 - \lambda)}{\lambda(1 - 2\lambda)^{\alpha}s^{\alpha}} \quad \text{or} \quad 1 = \frac{\alpha\delta(1 - \lambda)}{1 - 2\lambda}$$

Thus

$$1 - 2\lambda = \alpha\delta(1 - \lambda) \quad \text{and} \quad \lambda_c = \frac{1 - \alpha\delta}{2 - \alpha\delta}$$

The Cournot-Nash policy for this example is

$$g(s) = h(s) = \frac{1 - \alpha\delta}{2 - \alpha\delta}\, s$$

Note that the linearity of the Cournot-Nash policy function simplifies the response term, i.e., $h'(s) = \lambda$. Also note that

$$\lambda_M = 1 - \alpha\delta > \lambda_c = \frac{1 - \alpha\delta}{2 - \alpha\delta} \quad \text{and} \quad \lambda_M < 2\lambda_c$$

Hence for this example each country consumes less than the monopolists (or the cooperating countries) while *together* the duopolists consume more. The dynamics for this example is given by

$$
\begin{aligned}
x_{t+1} &= f(x_t) - g(f(x_t)) - h(f(x_t)) \\
&= f(x_t) - 2g(f(x_t)) \\
&= x_t^{\alpha}\left(\frac{\alpha\delta}{2 - \alpha\delta}\right)
\end{aligned}
$$

Hence for a steady state,

$$\bar{\bar{x}} - \bar{\bar{x}}^{\alpha}\left(\frac{\alpha\delta}{2 - \alpha\delta}\right) \qquad \text{or} \qquad \bar{\bar{x}} = \left(\frac{\alpha\delta}{2 - \alpha\delta}\right)^{\frac{1}{1-\alpha}}$$

This situation is depicted in Figure 5. Note that since under the Cournot-Nash solution both participants extract more fish as a function of the amount of fish than the monopolist, there will be less fish in the Cournot-Nash steady state than under monopoly steady state.

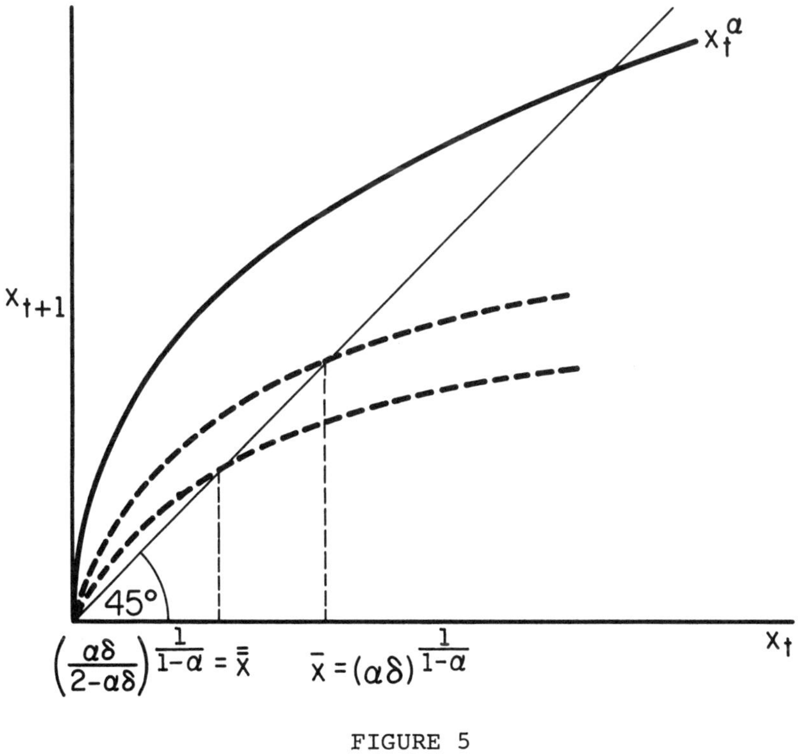

FIGURE 5

§4. PROFIT MAXIMIZATION UNDER MONOPOLY

In this and the next sections it will be easier to study the basic structure of the problem under the assumption of a linear growth function for fish. Suppose that the fish dynamics are given by

$$x_{t+1} = r\, x_t$$

where $r > 1$, and $\delta r > 1$. In this case it is profitable under monopoly to let the fish population increase. Suppose the country or firm maximizes the discounted sum of profits over time (again it is possible to think of two countries or firms maximizing a convex combination of their profits). Let

$$V(s) = \max_{\{q_t\}} \sum_{t=0}^{\infty} \delta^t\, \pi(q_t)$$

subject to

$$q_t + x_t = rx_{t-1} \qquad \text{and} \qquad q_0 + x_0 = s$$

Here $\pi(q_t)$ is the profit function. For simplicity it will be assumed that there are no costs and that $\pi(q_t) = q_t D(q_t)$ is a concave function. Here D is the demand function which is assumed fixed for each t. Note that the assumption $\pi'(0) = +\infty$ might not be realistic in this situation, however, we shall proceed under the implicit assumption that all maximizing occurs in the interior. As in the case of a single utility maximizer, the following analysis can be made precise. See, for example, Levhari, Michener, and Mirman [4]. In particular, it can be shown that V is concave.

Using a dynamic programming argument, it follows that

$$V(s) = \max_{q}\{qD(q) + \delta V(r(s - q))\} \tag{19}$$

Hence

$$qD'(q) + D(q) = \delta rV'(r(s - q)) \tag{20}$$

Here $qD'(q) + D(q)$ is the marginal revenue for this market (c.f., Figure 6). Note that (20) implies that for a static model, monopoly output is found from the equation

$$\pi'(q) = qD'(q) + D(q) = 0 \tag{21}$$

Hence if q_M satisfies (21), $V'(r(s - q_M)) = 0$. Let s_M be sustainable with output q_M, i.e., $s_M = rs_M - q_M$. Then for $s > s_M$ fish are no longer a scarce resource. For $s < s_M$, letting $q = q(s)$, equation 19 may be differentiated to yield

$$V'(s) = [qD'(q) + D(q)]q' + \delta rV'(r(s - q))(1 - q') \tag{22}$$

Using (20), it follows that

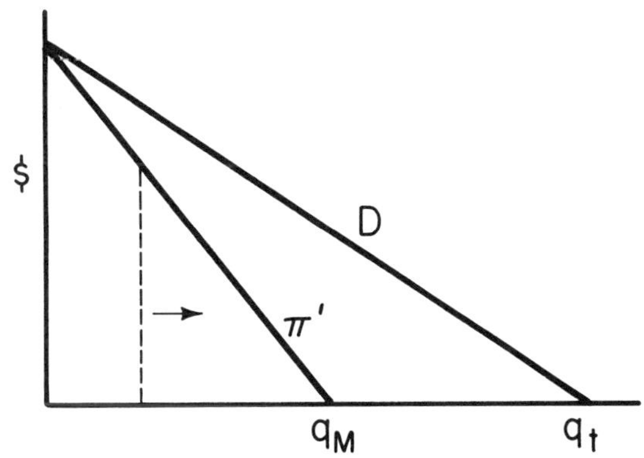

FIGURE 6

$$V'(s) = \delta r V'(r(s - q(s))) \tag{23}$$

This again yields the envelope equation

$$V'(s) = q(s)D'(q(s)) + D(q(s)) \tag{24}$$

Equation (24) implies that the value of an extra fish is merely the marginal revenue of selling the optimal amount of fish $q(s)$. Using the envelope equation (24) in (20), yields

$$\begin{aligned} q(s)D'(q(s)) + D(q(s)) = \delta r[&q(r(s - q(s))) \\ &\cdot D'(q(r(s - q(s)))) \\ &+ D(q(r(s - q(s))))] \end{aligned} \tag{25}$$

Recalling that $\pi'(q) = D'(q)q + D(q)$, (25) becomes

$$\pi'(q(s)) = \delta r \pi'(q(r(s - q(s))))$$

Note that π' is merely marginal revenue since costs have been assumed away. In general the discrete Euler equations are

$$\pi'(q(rx_t)) = \delta r \pi'(q(r(x_{t+1}))) \qquad t = 1, 2, \ldots \tag{26}$$

Since $\delta r > 1$, marginal revenue must be decreasing over time. This is illustrated in Figure 6. Hence under an optimal policy the population of fish grows over time. Note also that (26), which describes the optimal dynamics of the fish population, cannot have a natural steady state, i.e., where $\delta r = 1 (\delta f' = r\delta = 1)$, as in the utility maximization case since $\delta r \neq 1$. However, a market steady state can occur at the point at which a static equilibrium occurs; i.e., where $\pi'(q_M) = 0$, q_M is illustrated in Figure 6. The dynamics of

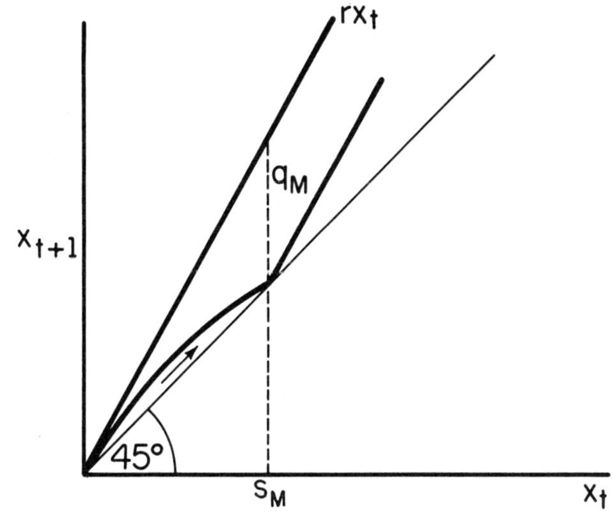

FIGURE 7

the process is depicted in Figure 7. The dynamics of this process is

$$x_{t+1} = rx_t - q(rx_t)$$

In the steady state

$$s_M = rs_M - q_M$$

so that

$$\frac{q_M}{r - 1} = s_M$$

Note that for $s < s_M$, $x_t \to s_M$, while for $s > s_M$, i.e., when the value of fish is zero, $x_t \to \infty$.

Next consider an example of this model. Suppose that $D(x) = \alpha - \beta x$, $\alpha, \beta > 0$, so that $D'(x) = -\beta$, and

$$\pi'(x) = \alpha - 2\beta x$$

Equation (26) becomes

$$\alpha - 2\beta q(s) = \delta r[\alpha - 2\beta q(r(s - q(s)))]$$

Letting $q(x) = A + Bx$, then

$$\alpha - 2\beta[A + Bs] = \delta r\{\alpha - 2\beta[A + Br(s - A - Bs)]\}$$

or

$$\alpha - 2\beta A - 2\beta Bs = \delta r[\alpha - 2\beta A - 2\beta Brs + 2\beta BrA + 2\beta B^2 rs]$$

Hence, solving for A and B, it follows that

$$A = \frac{\alpha(1 - \delta r)}{2\beta\delta r(r - 1)} < 0$$

and

$$0 < B = 1 - \frac{1}{\delta r^2} < 1$$

The consumption function $q(s)$ is depicted in Figure 8. Note that $\pi'(0) = \alpha$. Hence an interior maximum does not occur. However, corner solutions turn out to present no difficulty

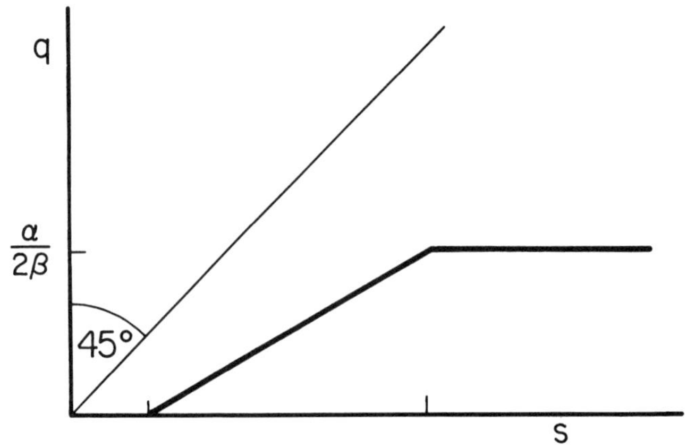

FIGURE 8

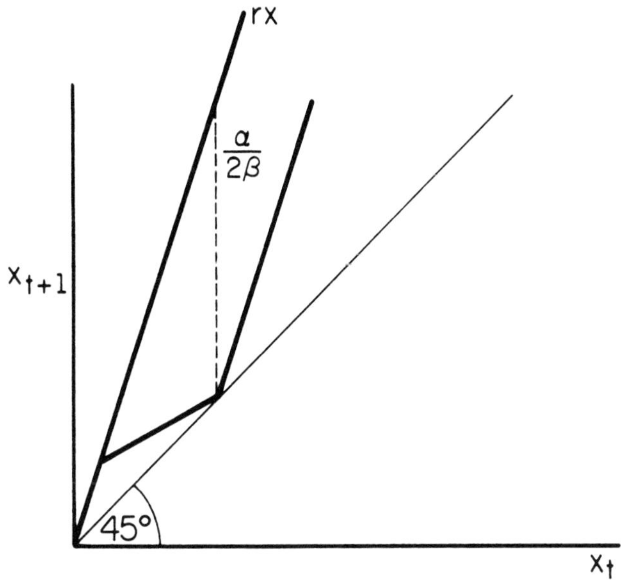

FIGURE 9

in this case since the stock of fish is never depleted. This is not the case when $\delta r < 1$. Thus when $\delta r < 1$ the analysis becomes considerably more difficult since Kuhn-Tucker type arguments have to be made. For this case, the dynamics are depicted in Figure 9.

§5. PROFIT MAXIMIZATION UNDER DUOPOLY

In this section, as in the previous section, the dynamics of the fish population is assumed to be linear. Also it is assumed that a static equilibrium in the Cournot sense is unique. Here again the objective of the firm is to maximize the sum of discounted profits given the biological constraints of the fish population as well as the behavior of the other firm or country as represented by the function h(s). Hence, let

$$V(s) = \max_{\{q_t\}} \sum_{t=0}^{\infty} \delta^t \pi(q_t)$$

subject to

$$q_t + x_t + h(rx_{t-1}) = rx_{t-1} \qquad t = 1, 2, \ldots$$

and

$$q_0 + x_0 + h(s) = s$$

As in the duopoly problem with agents maximizing the sum of discounted *utility*, the analysis in this section suffers from the difficulties that the function V(s) may not have all the requisite differentiability and concavity conditions used in this analysis. These problems will be illustrated in the next section and will be ignored in this section.

Using the dynamic programming formulation, it follows that

$$V(s) = \max_{q}\{D(q + h(s))q + \delta V(r(s - q - h(s)))\} \tag{27}$$

Notice that there are now two effects or externalities introduced by the duopoly problem. The utility maximization problem contained only one externality, namely, the other country affected only the dynamics of the fish population. Its policy had no effect, except indirectly, on the utility of the other duopolist. In this problem the other country or firm not only affects the dynamics of the fish population but also affects present, as well as future, profits through its effect on the price. This market externality is in fact the original externality studied by Cournot.[1]

Continuing the analysis with these externalities in mind, differentiation of the right hand side of (27) yields

$$D'(q + h(s))q + D(q + h(s)) = \delta rV'(r(s - q - h(s)) \quad (28)$$

Here $D'(q + h(s))q + D(q + h(s))$ is the marginal revenue for this firm given that the other firm's output is $h(s)$. A Cournot equilibrium occurs when $q = h(s)$ and

$$D'(2q_c)q_c + D(2q_c) = 0 \quad (29)$$

This is the *static* Cournot equilibrium for this market. As in the monopoly case at q_c, $V' = 0$. Hence if s_c is sustainable at output $2q_c$, i.e., $rs_c - 2q_c = s_c$, then $s > s_c$ implies that fish are extraneous and are no longer a scarce resource. While for $s < s_c$, using (28), and the policy $q = q(s)$,

$$\begin{aligned} V'(s) = {} & [D'(q(s) + h(s))q(s)(q'(s) + h'(s)) \\ & + D(q(s) + h(s))(q'(s))] + \delta rV'(r(s - q(s) \\ & - h(s)))(1 - h'(s) - q'(s)) \end{aligned} \quad (30)$$

Hence from (28)

$$\begin{aligned} V'(s) = {} & D'(q(s) + h(s))q(s)h'(s) \\ & + \delta rV'(r(s - q(s) - h(s)))(1 - h'(s)) \end{aligned} \quad (31)$$

---

[1]Actually, another externality should be introduced here. Namely, costs which have been omitted altogether should be introduced, especially costs which depend on the stock size of fish population. These cost externalities can be quite important. However, this complicates matters considerably and has been omitted.

It is at this point that the two externalities become clear. Suppose that there is an extra fish. The value to the duopolist is the sum of two terms. The first term $D'(q(s) + h(s)) q(s)h'(s)$ is the loss in revenue due to the fact that the other duopolist increases its supply and thus lowers the price. (Note that both duopolists increase supply but the envelope theorem takes account of the duopolist doing the maximization problem through (28)). The next term is a gain in future profits due to an increase in the fish population through the term $\delta r V'(r(s - g(s) - h(s)))$. This term is multiplied by $1 - h'(s)$ taking account of the dynamic externality or the fact that the other duopolist also catches some of the extra fish. This leads to the "envelope" result,

$$V'(s) = D'(q(s) + h(s))q(s)h'(s) + [D'(q(s) + h(s))q(s) + D(q(s) + h(s))](1 - h'(s)) \quad (32)$$

or

$$V'(s) = D(q(s) + h(s))(1 - h'(s)) + D'(q(s) + h(s))q(s) \quad (33)$$

Equation (33) may be interpreted as follows. The value to one of the duopolies of adding an extra fish is the increase in revenue due to selling his share of the extra fish, i.e., $1 - h'(s)$ minus the revenue lost due to the decreased price which results from the competitors increased sales.

Using the envelope relation in (28) yields

$$
\begin{aligned}
&D'(q(s) + h(s))q(s) + D(q(s) + h(s)) \\
&= \delta r\{[D(q(r(s - q(s) - h(s))) + h(r(s - q(s) - h(s)))) \\
&\quad + D'(q(r(s - q(s) - h(s))) + h(r(s - q(s) - h(s)))) \\
&\quad \cdot q(r(s - q(s) - h(s)))](1 - h'(r(s - q(s) - h(s))))] \\
&\quad + D'(q(r(s - q(s) - h(s))) + h(r(s - q(s) - h(s)))) \\
&\quad \cdot q(r(s - q(s) - h(s))))h'(r(s - q(s) - h(s)))\} \qquad (34)
\end{aligned}
$$

This expression can be written in general, but it would not really add very much to the description. On the other hand, a general expression may be written for the dynamics under a Cournot equilibrium. When $h(s) = q(s)$, using the residual marginal revenue function $\tilde{\pi}'(q) = D'(2q)q + D(2q)$, the general expression is

$$
\begin{aligned}
\tilde{\pi}'(q(s)) = \delta r\{&\tilde{\pi}'(q(r(s - 2q(s))))(1 - h'(r(s - 2q(s)))) \\
&+ D'(q(r(s - 2q(s))))q(r(s - 2q(s))) \\
&\cdot h'(r(s - 2q(s)))\} \qquad (35)
\end{aligned}
$$

For this problem a market steady state occurs where $\tilde{\pi}'(q_c) = 0$, for at this point $q' = 0$, i.e., at s_c fish are no longer scarce. Hence a static market equilibrium is again a dynamic market equilibrium. This situation is depicted in Figure 10, without the dynamics below the market equilibrium. Note that (35) implies that the Cournot solution has a stronger tendency to deplete the stock of fish than the cooperative solution. This is so since the factor $1 - h'$ multiplies the residual marginal revenue on the right hand side and there is, in addition, an extra term on the right hand side which is negative. Hence, even if

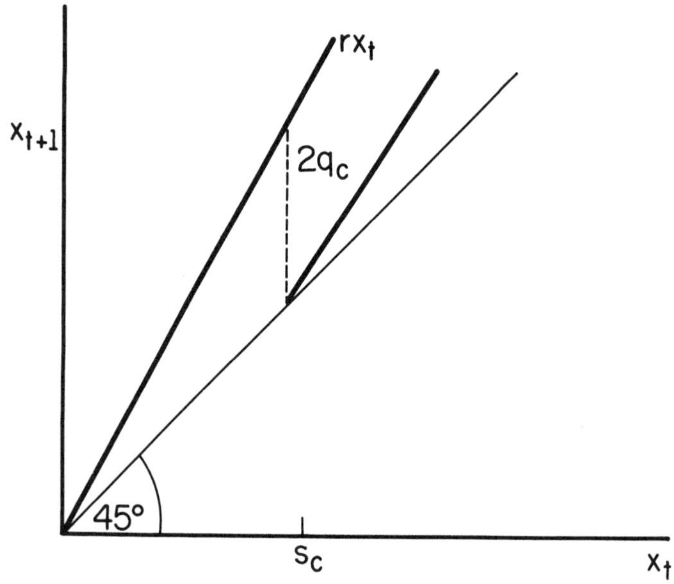

FIGURE 10

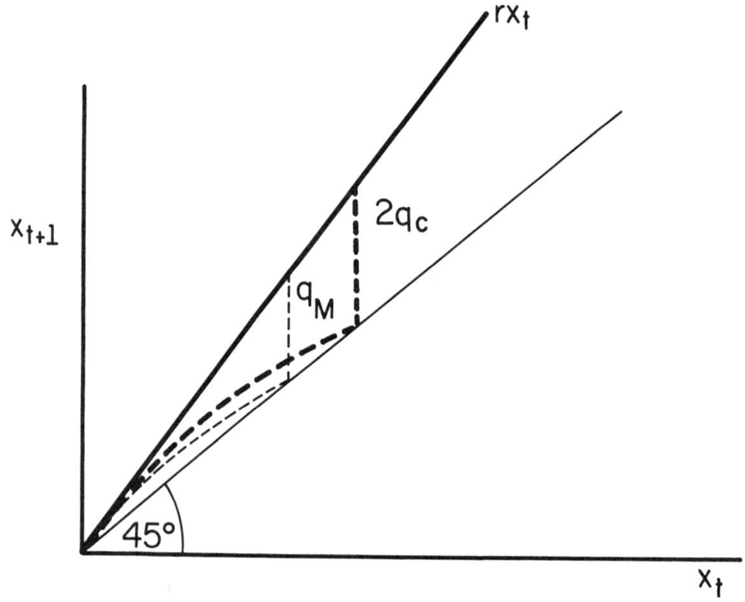

FIGURE 11

$\delta r > 1$ the residual marginal revenue on the right side could be greater than the residual marginal revenue on the left side, implying a declining stock of fish.

In comparing the monopoly case and the duopoly case, all that really can be said for sure is that at the duopoly equilibrium more fish is supplied to the market. A possible example of the dynamics is depicted in Figure 11.

An example similar to the one discussed in the last section is also applicable here. Again, as with the utility maximizing example, the reaction function h' is a constant. Even though the optimal policy is linear, this example is not as straightforward as the example of the last section. Letting $D(q) = \alpha - \beta q$, and $\tilde{\pi}'(q) = \alpha - 3\beta q$, equation (35) becomes

$$\begin{aligned}\alpha - 3\beta q(s) = \delta r\{&(\alpha - 3\beta q(r(s - 2q(s))) \\ &\cdot (1 - q'(r(s - 2q(s))))) \\ &- \beta q(r(s - 2q(s)))q'(r(s - 2q(s)))\}\end{aligned}$$

Letting $\bar{A} + \bar{B}s = q(s)$, then

$$\begin{aligned}\alpha - 3\beta[\bar{A} + \bar{B}s] = \delta r\{&(\alpha - 3\beta[\bar{A} + \bar{B}r(1 - 2B)s \\ &- \bar{B}r2\bar{A}])(1 - \bar{B}) \\ &- \beta[\bar{A} + \bar{B}r(1 - 2\bar{B})s - \bar{B}r2\bar{A}]\bar{B}\}\end{aligned}$$

Hence

$$\bar{B} = 1 - \frac{1}{2}\sqrt{1 + \frac{3}{\delta r^2}}$$

The expression for $\bar{A}$ is not so straightforward.

The consumption function is depicted in Figure 12 and the dynamics of the process is depicted in Figure 13.

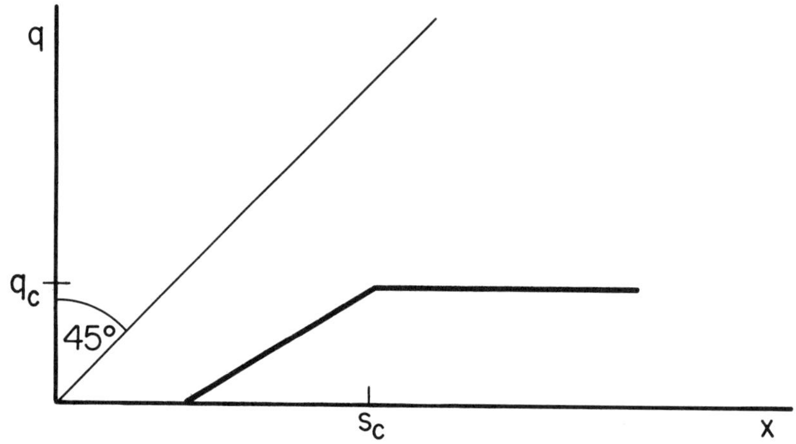

FIGURE 12

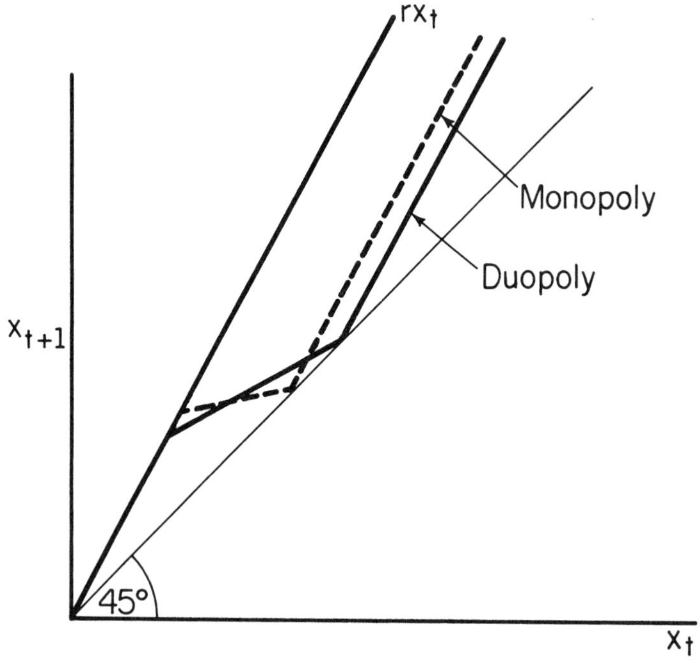

FIGURE 13

Note that the dynamics are

$$x_t - 2\bar{A} - 2\bar{B}x_t = x_{t+1}$$

and it can be shown that $2\bar{B} < B$. The duopoly dynamics can be compared to the monopoly dynamics, which is done in Figure 13. Note that the dynamics drawn in Fig. 13 represents only one possibility. Another possibility is that the monopoly might always extract less fish, i.e., the dotted line may lie wholly above the solid line. Finally, as pointed out in the last section, it is possible the Cournot duopolists will deplete the stock of fish even though $\delta r > 1$. The above analysis does not apply in that case since the functional equations are only valid when there is no depletion. Otherwise, a Kuhn-Tucker argument is needed, and in this case a linear function would not be optimal. Hence this example is only valid when the stock is not depleted.

§6. AN EXAMPLE SHOWING THE VALUE FUNCTION NEED NOT BE CONCAVE

In this section it will be shown, by an example, that the conditions required for the analysis of the Cournot-Nash equilibrium are not necessarily satisfied. In particular, the value function V need not be concave or differentiable. In fact, it may not even be continuous. However, two qualifications should be made. First, the rate of return equals one and future utilities are not discounted. These assumptions are made for ease in calculations. The second point is that the analysis is carried out using an iterative process. This is not the same technique used above. The aim of the example is to show that difficulties may arise rather than explore all the possibilities. It is clear from the example that even though the value function is not concave or continuous, a policy function of the Cournot-Nash type does exist. On the other hand it might be the case that there is a nonuniqueness problem involved. However, this question is not explored.

Let the production process be linear, i.e.,

$$x_{t+1} = x_t + y_t \qquad y_t = y \qquad t = 0, 1, \ldots$$

The utility function is $u(c) = \log c$. It is assumed that future utilities are not discounted. The objective is to derive a consumption function and use it to derive the value function. The same technique would work when the future is discounted, although it is simpler notationally not to discount. Finite horizon optimal policies are derived for finite horizon problems and the limit is taken as the horizon tends to infinity. It is assumed that everything is consumed, and shared equally, in the last period. This assumption has no effect since the fraction is easily factored out in the log utility case.

Consider a one period horizon problem

$$V_1(x) = \max_{\mu}\{\log \mu + \log(x + y - \mu - \nu)\}$$

Here ν is the policy of the other participant. The first order condition is

$$\frac{1}{\mu} = \frac{1}{x + y - \mu - \nu}$$

Since both participants are assumed to be exactly the same, μ may be found from the equation

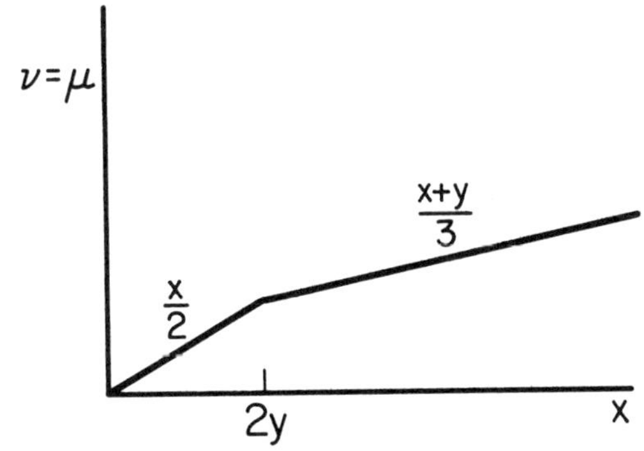

FIGURE 14

$$x + y = 3\mu$$

i.e.,

$$\mu = \frac{x + y}{3}$$

The policy is valid only if

$$2\mu = \frac{2(x + y)}{3} \leq x$$

i.e.,

$$y \leq \frac{x}{2} \qquad \text{or} \qquad x \leq 2y$$

Otherwise it is assumed each participant shares half of the fish so that if $x \leq 2y$, $\mu = x/2$. This policy function is illustrated in Fig. 14. Hence the one period horizon value function is

$$V_1(x) = \begin{cases} \log \frac{x}{2} + \log y = \log x + \log \frac{y}{2}, \text{ if } x \geq 2y \\ \log \frac{x+y}{3} + \log \frac{x+y}{3} = 2 \log(x+y) - \log 3^2, \text{ if } x \geq 2y \end{cases}$$

<u>Note 1</u> $V_1(x)$ is continuous at $x = 2y$ since $\log 2y + \log y/2 = 2 \log y$ and $2 \log(x + y) - \log 3^2 = 2 \log 3y - \mathrm{lcg}\ 3^2 = 2 \log y + \log 3^2 - \log 3^2 = 2 \log y$.

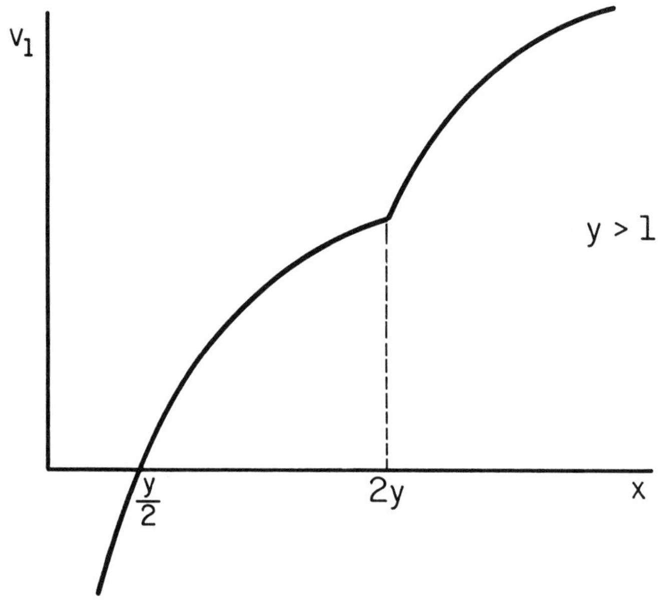

FIGURE 15

<u>Note 2</u> V_1 is not concave, since at $x = 2y$, $1/x = 1/2y < 2/x + y = 2/3y$. In fact, possible $V_1(x)$ are depicted in Figures 15 and 16.

Next, consider a two period horizon problem,

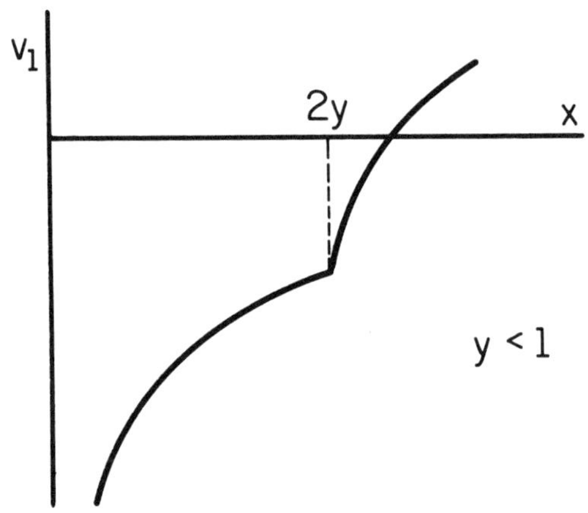

FIGURE 16

$$V_2(x) = \begin{cases} \max_\mu\{\log \mu + \log(x + y - \mu - \nu) + \log \frac{y}{2}\} \\ \qquad\qquad \text{if } x + y - \mu - \nu \le 2y \\ \max_\mu\{\log \mu + 2 \log(x + 2y - \mu - \nu) - \log 3^2\} \\ \qquad\qquad \text{if } x + y - \mu - \nu \ge 2y \end{cases}$$

Again the first order conditions in the first branch yields

$$\mu = \frac{x + y}{3}$$

which is valid whenever $x + y - 2\mu \le 2y$ or $x - 2\mu \le y$. Replacing μ, we have

$$x - \frac{2}{3}(x + y) \le y \qquad \text{or} \qquad \frac{x}{3} \le \frac{5}{3}\, y = x \le 5y$$

Note also that $\mu = (x + y)/3$ only if $x \ge 2y$, otherwise $\mu = x/2$. The second branch of $V_2(x)$ yields

$$\frac{1}{\mu} = \frac{2}{x + 2y - \nu - \mu}$$

Again, letting $\nu = \mu$ and solving for μ

$$\mu = \frac{x + 2y}{4}$$

which is valid whenever $x + y - 2\mu \ge 2y$; i.e.,

$$x + y - \frac{x + 2y}{2} \ge 2y$$

or $2(x + y) - x - 2y \ge 4y$, $x \ge 4y$. Note that these intervals overlap, i.e., the first branch is valid when $x \le 5y$ and the second when $x \ge 4y$. Hence in the interval $[4y, 5y]$ the greatest value of the two maximization problems of $V_2(x)$ will be chosen. This amounts to choosing one branch of the Nash solution. This is done as follows. The first branch with $\mu = (x + y)/3$ becomes

$$V_2^1(x) = 2 \log(x + y) - \log 3^2 + \log \frac{y}{2}$$

while the second branch with $\mu = (x + 2y)/y$ becomes

$$V_2^2(x) = 3 \log (x + 2y) - \log 3^2 4^2$$

Consider

$$V_2^1(x) - V_2^2(x) = 2 \log (x + y) - \log 3^2 + \log \frac{y}{2}$$

$$- 3 \log (x + 2y) + \log 3^2 4^2$$

$$= \log \frac{(x + y)^2 \, 8y}{(x + 2y)^3}$$

Let $x = \alpha y$, then

$$\log \frac{(x + y)^2 \, 8y}{(x + 2y)^3} = \log \frac{(\alpha + 1)^2 \, 8}{(\alpha + 2)^3}$$

Hence for $V_2^1(x) = V_2^2(x)$,

$$8(\alpha + 1)^2 = (\alpha + 2)^3$$

i.e.,

$$8\alpha^2 + 16\alpha + 8 = 8 + 12\alpha + 6\alpha^2 + \alpha^3$$

$$2\alpha^2 + 4\alpha = \alpha^3$$

So that

$$\alpha^2 = 2\alpha + 4$$

$$\alpha^2 - 2\alpha - 4 = 0$$

$$\alpha = \frac{2 \pm \sqrt{4 + 4(4)}}{2} = 1 \pm \sqrt{5} < 4$$

Thus $V_2^2(4y) > V_1^1(4y)$, and $V_2^2(x) > V_1^1(x)$ for all $x > 4y$. Hence the function V_2 may be used to derive a three period horizon optimal policy,

$$V_3(x) = \begin{cases} \max\limits_{\mu}\{\log \mu + \log (x + y - \nu - \mu) + \log \frac{4^2}{2^2}\} \\ \qquad \text{if } x + y - \mu - \nu \leq 2y \\ \max\limits_{\mu}\{\log \mu + 2 \log (x + 2y - \nu - \mu) + \log \frac{4}{2} - \log 3^2\} \\ \qquad \text{if } 2y \leq x + y - \mu - \nu \quad 4y \\ \max\limits_{\mu}\{\log \mu + 3 \log (x + 3y - \nu - \mu) - \log 3^2 4^2\} \\ \qquad \text{if } 4y \leq x + y - \mu - \nu \end{cases}$$

Here $V_2(x)$ is depicted in Figure 17, while the optimal policy is depicted in Figure 18. In the limit, it can be shown that the value function is a continuation of the value function shown in Figure 17. In general, the jumps in the value function occur at the points

$$x = 1 + \sum_{j=1}^{M} j$$

The general consumption function is also a continuation of the

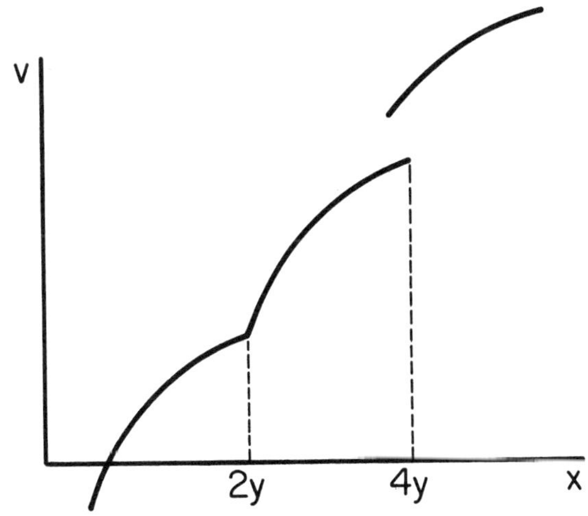

FIGURE 17

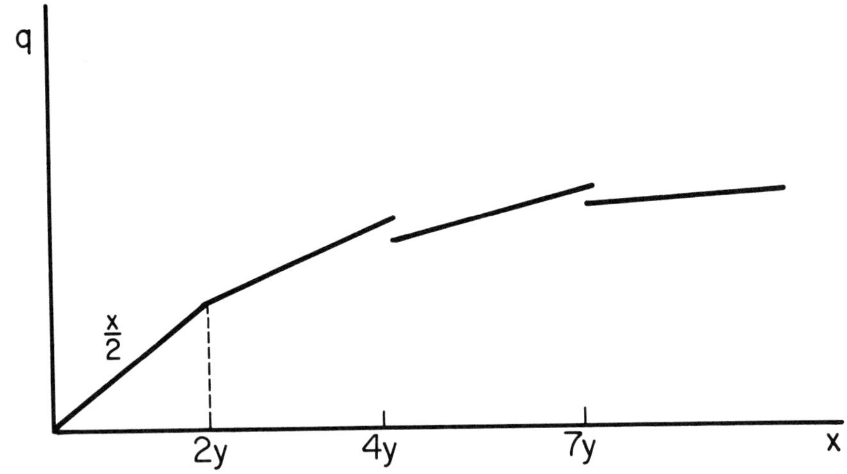

FIGURE 18

function depicted in Figure 18. The general slope of this consumption function is given by

$$\frac{x + (M - 2)y}{M}$$

ACKNOWLEDGMENT

Research support from the National Science Foundation under Grants SOC 76-11583 and SOC 77-27340 is gratefully acknowledged.

A BIBLIOGRAPHICAL NOTE

Much of the existing literature on the problems of fishing is contained in a book by Clark [1]. Also works by Clark [2] and Smith [6] are useful and interesting. The use of dynamic programming in the area of renewable resources is contained in Levhari, Michener, and Mirman [4]. However, none of the work cited above deals with Cournot or duopoly models.

Levhari and Mirman [3] and Takayama [7] contain work on duopoly models, but not duopoly models in which there are two externalities or in which the market externality has an effect on the optimal policy.

REFERENCES

1. Clark, C. W., Mathematical Bioeconomics, Wiley-Interscience, New York, 1976.
2. Clark, Colin W., Mathematical Models in the Economics of Renewable Resources, The Institute of Applied Mathematics and Statistics, Technical Report 77-4.
3. Levhari, David, and Mirman, Leonard J., The Great Fish War: An Example Using a Dynamic Cournot-Nash Solution, March 1977.
4. Levhari, David, Michener, Ron, and Mirman, Leonard J., Dynamic Programming Models of Fishing, June, 1978.
5. Mirman, Leonard, and Zilcha, Itzhak, On Optimal Growth Under Uncertainty, Journal of Economic Theory, Vol. 11, No. 3 (December 1975).
6. Smith, V. L., On Models of Commercial Fishing, J. Polit. Econ., 77, 181-198 (1969).
7. Takayama, T., Dynamic Theory of Fisheries Economics -- II; Differential Game Theoretic Approach (faculty working papers), University of Illinois, October, 1977.

OPTIMAL RESOURCE ALLOCATION AS A DIFFERENTIAL GAME

M. H. I. Dore

University of Saskatchewan
Saskatoon, Canada

ABSTRACT

The paper treats the problem of resource allocation as a two person zero-sum differential game that is played by a manager of a multi-divisional firm and a Central Planning Board (CPB). The game is played over a finite horizon in which the manager chooses an investment strategy in response to prices set by the CPB. The manager's strategy as well as that of the CPB is constrained within the framework of the closed Leontief model.

The manager maximizes his bonuses, which is a linear functional of the value of output, and the CPB minimizes the bonuses by an appropriate price strategy. With certain regularity assumptions, it is shown that an equilibrium strategy as well as a saddle point exist. The analysis is in terms of two complementary variational problems which are capable of being synthesized.

§1. INTRODUCTION

The published literature[1] on the design of allocation mechanisms is mainly static in character. Secondly, the allocation mechanisms are largely confined to economies of pure exchange, and if production is introduced this is done only in a formal manner. The present paper incorporates production in an essential way and attempts to consider the allocation problem in a dynamic context. However, this paper makes the same kind of convexity assumptions that have proved to be indispensable in the literature in order to preserve some desirable properties of the mechanisms such as global optimality and informational efficiency. The importance of the convexity assumptions has been highlighted recently by Mount and Reiter [5].

This paper treats the allocation problem as a differential game played by a manager of a multi-divisional firm and a Central Planning Board (CPB). The game is played over a finite horizon in which the manager chooses a strategy of investing in capital equipment and producing an output in response to prices set by the CPB. The strategies are subject to very reasonable constraints. In particular, the CPB's price setting rule is in the nature of an equilibrium condition. Prices are set in order to guide investment decisions, so that in equilibrium the rate of return in all divisions of the firm are equalized. Thus the CPB is not some arbitrary planner, but more of a theoretical artifice, like Walras' auctioneer.

---

[1]For recent surveys of this literature, see, for example, Hurwicz [4], Reiter [6], and Dore and Kaser [3].

In Section 2, the basic economic relationships are described and the main constraints of the differential game are stated. The game itself is formulated in Section 3, which is followed by a discussion of the existence of an equilibrium of the game. The existence of a saddle point is guaranteed by a theorem due to Berkovitz [1].

§2. THE BASIC ECONOMIC RELATIONSHIPS

The entire paper rests on basic equations that describe the closed Leontief model of an economy or a multidivisional firm.

Let A and B by $n \times n$ matrices that are nonsingular; A is the matrix of current input coefficients and B is the matrix of capital coefficients. In addition: x is an $n \times 1$ vector of gross outputs; y is an $n \times 1$ vector of final outputs; u is an $n \times 1$ vector of investment; and p is an $1 \times n$ vector of prices. The multidivisional firm's input-output structure is described by

$$(A + B)x + y = x \tag{1}$$

Its alternative description is given by

$$\begin{aligned} &Ax + \lambda Bx = x && (\lambda = 1 + r) \\ &Ax + Bx + rBx = x && (0 < r < 1) \\ &\quad\Rightarrow rBx = y && \end{aligned} \tag{2}$$

In terms of prices

$$p(A + B) + v = p \tag{3}$$

or

$$pA + \lambda pB = p \tag{4}$$

$$\Longrightarrow rpB = v$$

Using (1), we may write

$$[I - A - B]x = y \tag{5}$$

or

$$Dx = y \qquad (D = I - A - B]) \tag{6}$$

and

$$Bx = (I - A)x - y \tag{7}$$

Introducing time we may write (7) as an inequality:

$$B(x(t + 1) - x(t)) \leq (I - A)x(t)$$

Suppose that àn investment of u(t) this period enhances productive capacity next period. Letting u(t) be a piecewise smooth, bounded and nonnegative control vector, the inequality becomes

$$B(x(t + 1) - x(t)) = (I - A)x(t) - u(t)$$

which in continuous time is

$$\dot{x} = B^{-1}[I - A]x(t) - B^{-1}u(t) \tag{8}$$

Equation (8) shows how gross output changes as a result of investment.[2] Note that it is linear in x and u.

In terms of prices, there exists a similar inequality:

$$\dot{y}' \leq rp(t)B \tag{9}$$

[2]See Brody [2].

That is, the change in output cannot exceed some positive rate of return on capital. Again, we can introduce a piecewise smooth, bounded and nonnegative control vector of prices so that (9) becomes

$$\dot{y}' = rp(t)B - p(t) \tag{10}$$

If we differentiate (6) with respect to time and substitute into (10), we have

$$(D\dot{x})' = rp(t)B - p(t)$$

and

$$\dot{x}' = rp(t)B(D^{-1})' - p(t)(D^{-1})' \tag{11}$$

It will become clear below that u(t) or investment is the manager's control vector and p(t) is the CPB's control vector. Section 3 uses the basic technology of this section to formulate a two-person zero-sum differential game.

§3. THE DIFFERENTIAL GAME

Suppose there exists one multidivisional firm with a manager at the head of this firm. The manager wishes to maximize his bonuses over the period $[t_0, T]$ by choosing an investment strategy $u^*(t_0), \ldots, u^*(T)$. His bonus is a linear function of the value of output and so he maximizes the latter at the prices set by the CPB. On the other hand, the CPB chooses a price strategy $p(t_0), \ldots, p(T)$ that minimizes the sum of bonuses over the same plan period. It is shown that a saddle point of the game is well defined, since for any arbitrarily given initial conditions a unique pair (u^*, p^*) for $t_0 \le t \le T$ exists. The game is formalized as the manager's problem, and the CPB's problem.

§3.1 The Manager's Problem

The manager's problem is to maximize his bonuses by an appropriate choice of u for all $t_0 \le t \le T$.

$$J[u^*, p] = \max \int_{t_0}^{T} [p(t)y(t)]dt \tag{12}$$

subject to

$$\dot{x} = B^{-1}[I - A]x(t) - B^{-1}u(t) \tag{8}$$

$$0 \le u(t) \le \int_{-\infty}^{t} y(\tau)d\tau \tag{13}$$

$$x(t_0) = x_0 \tag{14}$$

$$x(T) = x_T \tag{15}$$

Assuming that the manager can take p(t) as a parameter, substitute for y from equation (1) into equation (12):

$$J[u^*, p] = \max \int_{t_0}^{T} [p(t)[(I - A)x(t) - Bx(t)]]dt \tag{16}$$

Form the Hamiltonian,

$$\begin{aligned} H(x,u,p,\psi,t) &= p(t)[I - A]x(t) - p(t)Bx(t) \\ &\quad + \psi(t)[B^{-1}(I - A)x(t) - B^{-1}u(t)] \end{aligned} \tag{17}$$

Noting that x is a function of u,

$$\frac{\partial H}{\partial u} = p(t)[I - A]\frac{\partial x}{\partial u} - p(t)B\frac{\partial x}{\partial u}$$

$$+ \psi(t)B^{-1}[I - A]\frac{\partial x}{\partial u} - \psi(t)B^{-1} = 0$$

$$= p(t)D\frac{\partial x}{\partial u} + \psi(t)B^{-1}[1 - A]\frac{\partial x}{\partial u} - \psi(t)B^{-1} = 0 \qquad (18)$$

for all $t_0 \leq t \leq T$.

It is shown in the appendix that

$$\frac{\partial x}{\partial u} = D^{-1} + O(\varepsilon) \qquad \text{and that} \qquad \frac{\partial u}{\partial x} = D$$

which means that equation (18) can be used to solve for $\psi(t)$ for all $t_0 \leq t \leq T$:

$$\psi^*(t) = -p(t)[(I - A)D^{-1} - I]^{-1}B + O(\varepsilon)$$

$$= p(t)[I - (I - A)D^{-1}]^{-1}B + O(\varepsilon)$$

$$= p(t)(I + G + G^2 + \cdots]B + O(\varepsilon)$$

$$\psi^*(t) = p(t)FB + O(\varepsilon)$$

where

$$G = (I - A)D^{-1} \qquad \text{and} \qquad F = I + G + G^2 \cdots \qquad (19)$$

Thus taking p(t) as given, the manager can solve for $\psi^*(t)$ for all $t_0 \leq t \leq T$, and having done this he can solve for his optimal controls $u^*(t)$:

$$\frac{\partial H}{\partial x} = p(t)(I - A) - p(t)B + \psi^*(t)[B^{-1}(I - A) - B^{-1}\frac{\partial u}{\partial x}]$$

$$= -\dot{\psi}(t)$$

or

$$[p(t) + \psi^*(t)B^{-1}][I - A] - p(t)B + \dot{\psi}(t) = \dot{\psi}^*(t)B^{-1}\frac{\partial u}{\partial x}$$

and eliminating $\psi^*(t)$, we have

$$p(t)(I - A) + p(t)F(I - A) - p(t)B + \dot{p}(t)FB$$
$$= p(t)FD + O(\varepsilon) \qquad \text{for all } t_0 \le t \le T \tag{20}$$

Now differentiate both sides of (20) partially with respect to p and multiply through by x(t):

$$(I - A)x(t) - Bx(t) + F(I - A)x(t) + \frac{\partial \dot{p}}{\partial p} FBx(t)$$
$$= FDx(t) + O(\varepsilon)$$

However, $\dot{p}$ is independent of p, since the time rate of change of prices cannot possibly be affected by the level of prices. Hence $\partial\dot{p}/\partial p = 0$, which means that, using equation (5), it reduces the above equation to

$$y(t) + F[y(t) + Bx(t)] = FDx(t) + O(\varepsilon)$$

or

$$x(t) = -[FB - FD]^{-1}[I + F]y(t) + O(\varepsilon)$$

Finally, since equation (8) is also a necessary condition for the optimum,

$$[I - A]^{-1}[B\dot{x} + u(t)] = -[FB - FD]^{-1}[I + F]y(t) + O(\varepsilon)$$

$$u^*(t) = -[I - A][FB - FD]^{-1}[I + F]y(t) - B\dot{x} + O(\varepsilon)$$

and substituting equation (11), we get

$$u^*(t) = -[I - A][FB - FD]^{-1}[I + F]y(t)$$
$$- B[rD^{-1}B'p'(t) - D^{-1}p'(t)] \tag{21}$$

Thus, to within a first order approximation, equation (21) shows that the manager's optimal controls are a function of his output, and prices set by the CPB.

It is obvious that the above conditions are necessary and sufficient for a maximum, since the constraint (8) is linear and the objection function is quadratic.[3] Note, too, that differentiating H(·) with respect to ψ yields the constraint (8).

§3.2 The CPB's Problem

The CPB's problem, as stated above, is to minimize the sum of bonuses. It therefore minimizes the *value* of output with an appropriate choice of p for all $t_0 \le t \le T$; i.e.,

$$J[u,p^*] = \min \int_{t_0}^{T} [-p(t)[(I - A)x(t) - Bx(t)]]dt \qquad (22)$$

subject to

$$\dot{x}' = rp(t)B(D^{-1})' - p(t)(D^{-1})' \qquad (11)$$

$$0 \le p(t) \le \eta(t)$$

where each $\eta_i(t)$ is some positive constant, and $x(t_0) = x_0$, $x(T) = x_T$. Form the Hamiltonian

$$\bar{H}(x,p,\xi,t) = p(t)Bx(t) - p(t)(I - A)x(t) + \xi(t)[rp(t)B(D^{-1})' - p(t)(D^{-1})'] \qquad (23)$$

$$\frac{\partial \bar{H}}{\partial p} = Bx(t) - (I - A)x(t) + \xi(t)[rB(D^{-1})' - (D^{-1})'] = 0$$

So

$$\xi^*(t) = [(I - A)x(t) - Bx(t)][rB - I)(D^{-1})']^{-1}$$

$$\xi^*(t) = y(t)D'[rB - I]^{-1} \quad \text{for all } t_0 \le t \le T \qquad (24)$$

---

[3]From equation (6), $y = Q_1x$ (say) and from equation (11), $p = Q_2x$ so that $py = x'Q_2Q_1x$, or $py = x'\tilde{Q}x$

$$\frac{\partial \bar{H}}{\partial x} = p(t)B - p(t)[I - A] = -\dot{\xi}(t)$$

$$p(t)[A + B - I] = -\dot{\xi}(t)$$

Multiply through by -1 and we have, for all $t_0 \leq t \leq T$,

$$p(t)D = \dot{\xi}(t) \tag{25}$$

First, differentiate equation (24) with respect to time and substitute into (25):

$$p^*(t)D = \dot{y}(t)D'[rB - I]^{-1}$$

$$p^*(t) = \dot{y}(t)D'[rB - I]^{-1}D^{-1} \tag{26}$$

Now differentiate (6) with respect to time and substitute into (26):

$$p^*(t) = D\dot{x}D'(rB - I)^{-1}D^{-1}$$

and, substituting equation 8, we may write

$$p^*(t) = D[B^{-1}(I - A)x^*(t) - B^{-1}u^*(t)]D'[rB - I]^{-1}D^{-1}$$

$$\text{for all } t_0 \leq t \leq T \tag{27}$$

Thus the prices, which are the CPB's controls, are a function of the current value of the state vector x and the manager's controls u*, for all $t_0 \leq t \leq T$.

§4. EXISTENCE OF AN EQUILIBRIUM

In Section 3 the differential game was developed as two complementary control problems. However, a synthesis is quite straightforward due to the theoretical developments of Berkovitz [1]. This is outlined below.

In the previous section it was shown that both in the manager's problem and in the CPB's problem the costate variables can be eliminated. This reduced each problem to a single equation. The manager's problem is reduced to equation (21) and that of the CPB to equation (27). Thus an equilibrium of the differential game requires the simultaneous solution of equations (21) and (27) for all $t_0 \le t \le T$. Therefore, there exists a unique trajectory of (u^*, p^*), for all $t_0 \le t \le T$.

Having established the existence of (u^*, p^*), it is easy to substitute into equations (16) and (22) and obtain a saddle point $H[x,u^*,p^*,t]$ such that

$$H[x,u,p^*,t] \le H[x,u^*,p^*,t] \le H[x,u^*,p,t]$$

$$\text{for all } t_0 \le t \le T$$

In fact, for the existence of the saddle point, we could rely on a theorem of Berkovitz. Note that the manager's objective functional (equation (16)) is concave,[4] and the CPB's is convex. As for the two constraints (8) and (11), first transpose both sides of equation (11) and add the result to equation (8). We then get

$$\dot{x} = \frac{1}{2}[B^{-1}(I - A)x(t) - B^{-1}u(t) + rD^{-1}B'p'(t) - D^{-1}p'(t)]$$

which is separable in u and p. If these conditions are met, then according to Berkovitz's Theorem 7, Corollary 1, a saddle point is guaranteed.

---

[4] See footnote 3.

§5. CONCLUSION

This paper is based on the fundamental equations of the Leontief model. The objective functionals of the two complementary control problems and the constraints of equations (8) and (11) rest on the Leontief model. The economic validity of these two equations is, of course, crucial. The meaning of (8) is simple: an investment in time period 1 alters the capital structure (viz., matrix B) in period 2, which changes gross output x in period 2. In this sense, B is clearly time dependent, but since it is *completely* determined by u, it can be treated as constant. Note, too, that A, the matrix of current input coefficients is also determined by B. A and B together determine r (see equation (2) or (4)). Hence, r too is time dependent in the same way as B is time dependent. But the same trick works again.

In economic terms, equation (11) is more than just a "constraint": it is a constraint which is in the nature of an equilibrium condition. The powers of the CPB are not arbitrary; the value of increased output must be attributed according to the von Neumann-Sraffa rule, which requires that the rate of profit in all sectors be equalized. Under certain conditions this is equivalent to the equalization of marginal returns to new investment in all sectors.

Since the CPB in this model "obeys" certain theoretical propositions of economics, it is in a sense a dynamic analog of Walras' auctioneer. Therefore, the differential game formulation presented here can be regarded as a dynamic pseudo-tatônnement allocation mechanism with endogenous production.

§6. APPENDIX

Differentiating equation (8) partially with respect to u, we get

$$\frac{\partial \dot{x}}{\partial u} = B^{-1}(I - A)\frac{\partial x}{\partial u} - B^{-1}$$

or

$$\left.\frac{\partial x}{\partial u}\right|_{u<y} = (I - A)^{-1}\left(B \frac{\partial \dot{x}}{\partial u} + I\right) \qquad (A.1)$$

However, equation (1) can be written as

$$x = [I - A - B]^{-1}y$$

and

$$\frac{dx}{dy} = [I - A - B]^{-1} \qquad (A.2)$$

But

$$\lim_{\substack{u \to y \\ t_0 \le t \le T}} \left(\frac{\partial x}{\partial u}\right) = \frac{dx}{dy} = [I - A - B]^{-1} = D^{-1} \qquad (A.3)$$

Therefore, to first order approximation, we may write

$$\frac{\partial x}{\partial u} = D^{-1} + O(\varepsilon) \qquad \varepsilon \to 0 \text{ as } u \to y$$

Finally, there is need for a simple expression of $\partial u/\partial x$. Of course, if D were diagonal this would be easy, but in general D is not a diagonal matrix. However, it would not be unreasonable to suppose that $\partial \dot{x}/\partial x$ is diagonal. For the purpose of this exercise the diagonal constant can be taken as unity; thus in terms of components

$$\frac{\partial \dot{x}_i}{\partial x_j} = \delta_{ij} \qquad \text{(Kronecker delta)}$$

In economic terms this assumption amounts to ignoring the indirect effects.

Now rearrange equation (8) and differentiate partially with respect to x:

$$\frac{\partial u}{\partial x} = (I - A) - B\frac{\partial \dot{x}}{\partial x}$$

Utilizing the assumption made above,

$$\frac{\partial u}{\partial x} = (I - A - B) = D \qquad (A.4)$$

ACKNOWLEDGMENT

I would like to thank Bill Bishopp, Madan Gupta, and John Stephenson, all of the University of Saskatchewan, for helpful comments on earlier drafts of this paper.

REFERENCES

1. Berkovitz, L. D., A Variational Approach to Differential Games. In Advances in Game Theory, M. Dresher, L. S. Shapley, and A. W. Tucker (eds.), Annals of Mathematical Studies, No. 52, Princeton University Press, Princeton, N. J., 1964, pp. 127-174.

2. Brody, A., Optimal and time-optimal paths of the economy. In Contributions to Input-Output Analysis, A. P. Carter and A. Brody (eds.), Proceedings of the Fourth International Conference on Input-Output Techniques, Geneva, 8-12 January 1968, Volume 1, North-Holland, Amsterdam, 1970, pp. 62-76

3. Dore, M. H. I., and Kaser, M. C., The Millions of Equations Debate: Seventy Years After Barone (to be published).

4. Hurwicz, L., The Design of Mechanisms for Resource Allocation, American Economic Review, Vol. 63, 1-30 (1973)

5. Mount, K., and Reiter, S., Economic environments for which there are Pareto satisfactory mechanisms, Econometrica, Vol. 45, No. 4, 821-842 (May 1977).

6. Reiter, S., Information and performance in the (New)[2] Welfare Economics, The American Economic Review, Papers and Proceedings, Vol. 67, No. 1, 226-34 (February 1977).

A BILINEAR-QUADRATIC DIFFERENTIAL GAME IN ADVERTISING

K. Deal, S. P. Sethi, and G. L. Thompson

McMaster University, Hamilton, Ontario, Canada
University of Toronto, Toronto, Ontario, Canada
and
Graduate School of Industrial Administration
Carnegie-Mellon University, Pittsburgh, Pennsylvania

ABSTRACT

A duopolistic extension of the Vidale-Wolfe advertising model is formulated as a problem in differential games. An important feature of the problem is the presence of reaction terms driven by the difference in the advertising expenditures of the two duopolistic firms under consideration. Formula for the Nash turnpike is derived using the maximum principle. Several competitive encounters with different parameters of the model are numerically solved. Tentative conclusions are drawn by analyzing these results.

§1. INTRODUCTION

As one of the earliest management science applications to marketing, Vidale and Wolfe [1] developed a simple model of sales response to advertising that was consistent with their experimental observations. They argued that changes in rate of sales of a product depend on two effects, response to advertising which acts (via the response constant b) on the unsold portion of the market, and loss due to forgetting which acts (via the decay constant a) on the sold portion of the market. Thus

$$\dot{x} = bu(1 - x) - ax \qquad x(0) = x_0 \tag{1.1}$$

where x is the market share (i.e., the rate of sales expressed as a fraction of the market potential or saturation level) and u is the rate of advertising expenditure (a control variable) at time t.

Whereas Vidale and Wolfe offered their model primarily as a description of actual market phenomena represented by cases which they had observed, Sethi [2,3] formulated an optimal advertising model with the Vidale-Wolfe model as its dynamics. He obtained complete solutions for many variants of the optimal control problem so formulated (see also Sethi [4]).

Deal and Zionts [5] and Deal [6] offered a duopolistic extension of the Vidale-Wolfe advertising model and treated the problem by the differential game approach. With indices 1 and 2 denoting the two duopolists, respectively, we can describe their dynamics as follows:

$$\begin{aligned} \dot{x}_1 &= b_1u_1(1 - x_1 - x_2) - a_1x_1 \qquad x_1(0) = x_{10} \\ \dot{x}_2 &= b_2u_2(1 - x_1 - x_2) - a_2x_2 \qquad x_2(0) = x_{20} \end{aligned} \tag{1.2}$$

The main problem with their dynamics is the absence of a competitive term, i.e., in their model, the competitor's advertising expenditure has no direct effect on a firm's market share. This drawback has led us to the modified differential game model to be described in the next section.

The plan of this paper is as follows. Section 2 contains our formulation of the bilinear quadratic differential game. In Section 3 we apply the appropriate maximum principle [7] to obtain necessary optimality conditions. Section 4 deals with the analysis of the infinite horizon version of the problem and obtains the steady state Nash equilibrium solution, which we term as *Nash turnpike*. Computational analysis of finite horizon versions of the problem is presented in Section 5. These problems are numerically solved by applying a two-point-boundary-value algorithm. We perform sensitivity analysis of the answers by varying the parameters of the problem. Section 6 concludes the paper.

§2. THE BILINEAR QUADRATIC MODEL

The state equations of the bilinear model are modifications of (1.2). They are

$$\begin{aligned} \dot{x}_1 &= b_1u_1(1 - x_1 - x_2) + e_1(u_1 - u_2)(x_1 + x_2) \\ &\quad - a_1x_1 \qquad\qquad x_1(0) = x_{10} \\ \dot{x}_2 &= b_2u_2(1 - x_1 - x_2) + e_2(u_2 - u_1)(x_1 + x_2) \\ &\quad - a_2x_2 \qquad\qquad x_2(0) = x_{20} \end{aligned} \tag{2.1}$$

Subject to these equations, firm 1 wants to maximize

$$J_1 = w_1e^{-\rho T}x_1(T) + \int_0^T (c_1x_1 - u_1^2)e^{-\rho t}\, dt \tag{2.2}$$

while firm 2 wants to maximize

$$J_2 = w_2e^{-\rho T}x_2(T) + \int_0^T (c_2x_2 - u_2^2)e^{-\rho t}\, dt \tag{2.3}$$

where ρ is the constant discount rate, c_1, c_2, w_1, w_2 are constants, and T is the horizon time. The problem defined by (2.1)-(2.3) is a non-zero sum differential game (see [7,8]).

The salient feature of this model is that it contains reaction terms of the form $e_1(u_1 - u_2)(x_1 + x_2)$ and $e_2(u_2 - u_1)(x_1 + x_2)$. To explain these terms assume $e_1 = e_2$ and $u_1 > u_2$. The term $e_1(u_1 - u_2)x_2$ represents the persons who switched from firm 2 to firm 1 because of the excess advertising of firm 1; the term $e_1(u_1 - u_2)x_1$ is an approximation (and simplification) of the Ozga type term $e_1(u_1 - u_2)x_1x_2$ representing the number of people switching from firm 2 to firm 1 as a result of the indirect or word of mouth advertising stimulated by the excess advertising of firm 1. Since the total number of people going from firm 2 to firm 1 is $e_1(u_1 - u_2)(x_1 + x_2)$, these people must be subtracted from the second state equation. A similar analysis works when $u_1 < u_2$. Another reason for making these assumptions is to end up with a simple bilinear model.

§3. APPLICATION OF THE MAXIMUM PRINCIPLE

To apply the maximum principle, we formulate a Hamiltonian for each player as follows:

$$\begin{aligned} H_1 = {} & c_1x_1 - u_1^2 + \lambda_1[b_1u_1(1 - x_1 - x_2) \\ & + e_1(u_1 - u_2)(x_1 + x_2) - a_1x_1] \\ & + \lambda_2[b_2u_2(1 - x_1 - x_2) + e_2(u_2 - u_1)(x_1 + x_2) \\ & - a_2x_2] \end{aligned} \tag{3.1}$$

$$\begin{aligned} H_2 = c_2x_2 - u_2^2 &+ \gamma_1[b_1u_1(1 - x_1 - x_2) \\ &+ e_1(u_1 - u_2)(x_1 + x_2) - a_1x_1] \\ &+ \gamma_2[b_2u_2(1 - x_1 - x_2) + e_2(u_2 - u_1)(x_1 + x_2) \\ &- a_2x_2] \end{aligned} \tag{3.2}$$

where the current-value adjoint variables λ_1 and λ_2 and γ_1 and γ_2 satisfy the following differential equations:

$$\begin{aligned} \dot{\lambda}_1 = \rho\lambda_1 - \frac{\partial H_1}{\partial x_1} &= -c_1 + \lambda_1[\rho + a_1 + b_1u_1 - e_1(u_1 - u_2)] \\ &+ \lambda_2[b_2u_2 - e_2(u_2 - u_1)] \end{aligned} \tag{3.3}$$

$$\begin{aligned} \dot{\lambda}_2 = \rho\lambda_2 - \frac{\partial H_1}{\partial x_2} &= \lambda_1[b_1u_1 - e_1(u_1 - u_2)] \\ &+ \lambda_2[\rho + a_2 + b_2u_2 - e_2(u_2 - u_1)] \end{aligned} \tag{3.4}$$

$$\begin{aligned} \dot{\gamma}_1 = \rho\gamma_1 - \frac{\partial H_2}{\partial x_1} &= \gamma_1[\rho + a_1 + b_1u_1 - e_1(u_1 - u_2)] \\ &+ \gamma_2[b_2u_2 - e_2(u_2 - u_1)] \end{aligned} \tag{3.5}$$

$$\begin{aligned} \dot{\gamma}_2 = \rho\gamma_2 - \frac{\partial H_2}{\partial x_2} &= -c_2 + \gamma_1[b_1u_1 - e_1(u_1 - u_2)] \\ &+ \gamma_2[\rho + a_2 + b_2u_2 - e_2(u_2 - u_1)] \end{aligned} \tag{3.6}$$

Also, these adjoint variables satisfy the transversality conditions:

$$\begin{aligned} \lambda_1(T) &= w_1 \qquad & \lambda_2(T) &= 0 \\ \gamma_1(T) &= 0 \qquad & \gamma_2(T) &= w_2 \end{aligned} \tag{3.7}$$

Since the Hamiltonian maximizing conditions are necessary and sufficient for the bilinear quadratic problem (see [7]), we obtain the form of the optimal controls as:

$$\frac{\partial H_1}{\partial u_1} = 0 \Rightarrow u_1 = \frac{1}{2}[\lambda_1\{b_1(1 - x_1 - x_2) + e_1(x_1 + x_2)\} - \lambda_2 e_2(x_1 - x_2)] \tag{3.8}$$

$$\frac{\partial H_2}{\partial u_2} = 0 \Rightarrow u_2 = \frac{1}{2}[\gamma_2(b_2(1 - x_1 - x_2) + e_2(x_1 + x_2)) - \gamma_1 e_1(x_1 + x_2)] \tag{3.9}$$

If we substitute (3.8) and (3.9) into (2.1) and (3.3)-(3.7), we obtain the two-point-boundary-value problem, which when solved, yields optimal advertising for the two firms. This will be carried out in Section 5. In the next section, we take up the analysis of the infinite horizon problem and obtain the expression for the *Nash turnpike*.

§4. DERIVATION OF THE NASH TURNPIKE

For this derivation we assume $T = \infty$ and modify the objective functions of (2.2) and (2.3) as follows:

$$J_1 = \int_0^\infty (c_1 x_1 - u_1^2)e^{-\rho t}\, dt \tag{4.1}$$

$$J_2 = \int_0^\infty (c_2 x_2 - u_2^2)e^{-\rho t}\, dt \tag{4.2}$$

The differential game problem involves firm 1 trying to maximize (4.1) and firm 2 trying to maximize (4.2) subject to conditions (2.1).

To determine the Nash turnpike we set the right hand sides of (2.1) and (3.3)-(3.6) equal to zero and solve these six equations along with (3.8) and (3.9). This gives a system of eight bilinear equations in eight variables. We now sketch the reduction of this system to two simultaneous cubic equations in u_1 and u_2. However, one equation has no term in u_1^3 and the other has no term in u_2^3. It is possible, therefore, to solve these equations by an iterative application of the quadratic formula.

The eight equations to be solved are:

$$-a_1x_1 + b_1u_1(1 - x_1 - x_2) + e_1(u_1 - u_2)(x_1 + x_2) = 0 \tag{4.3}$$

$$-a_2x_2 + b_2u_2(1 - x_1 - x_2) + e_2(u_1 - u_2)(x_1 + x_2) = 0 \tag{4.4}$$

$$-c_1 + \lambda_1[\rho + a_1 + b_1u_1 - e_1(u_1 - u_2)] + \lambda_2[b_2u_2 - e_2(u_2 - u_1)] = 0 \tag{4.5}$$

$$\lambda_2[\rho + a_2 + b_2u_2 = e_2(u_2 - u_1)] + \gamma_1[b_1u_1 - e_1(u_1 - u_2)] = 0 \tag{4.6}$$

$$\gamma_1[\rho + a_1 + b_1u_1 - e_1(u_1 - u_2)] + \gamma_2[b_2u_2 - e_2(u_2 - u_1)] = 0 \tag{4.7}$$

$$-c_2 + \gamma_2[\rho + a_2 + b_2u_2 - e_2(u_2 - u_1)] + \gamma_1[b_1u_1 - e_1(u_1 - u_2)] = 0 \tag{4.8}$$

$$2u_1 = \lambda_1[b_1(1 - x_1 - x_2) + e_1(x_1 + x_2)] - \lambda_2e_2(x_1 + x_2) \tag{4.9}$$

$$2u_2 = \gamma_2[b_2(1 - x_1 - x_2) + e_2(x_1 + x_2)] - \gamma_1e_1(x_1 + x_2) \tag{4.10}$$

Subtracting (4.6) from (4.5) and (4.8) from (4.7), we obtain

$$\lambda_1 = \frac{c_1 + \lambda_2(a_2 + \rho)}{a_1 + \rho} \tag{4.11}$$

$$\gamma_2 = \frac{c_2 + \gamma_1(a_1 + \rho)}{a_2 + \rho} \tag{4.12}$$

Substituting (4.11) into (4.6) and (4.12) into (4.7) yields

$$\lambda_2 = \frac{c_1[e_1(u_1 - u_2) - b_1u_1]}{D(\rho)} \tag{4.13}$$

$$\gamma_1 = \frac{c_2[e_2(u_2 - u_1) - b_2u_2]}{D(\rho)} \tag{4.14}$$

where

$$\begin{aligned} D(\rho) = {} & (a_1 + \rho)(a_2 + \rho) + (a_1 + \rho)b_2u_2 + (a_2 + \rho)b_1u_1 \\ & + (u_1 - u_2)[(a_1 + \rho)e_2 - (a_2 + \rho)e_1] \end{aligned} \tag{4.15}$$

Multiplying (4.3) by a_2 and (4.4) by a_1 and adding gives

$$x_1 + x_2 = \frac{a_2b_1u_1 + a_1b_2u_2}{D(0)} \tag{4.16}$$

Putting (4.11), (4.13), and (4.16) into (4.9) and multiplying by $D(0)D(\rho)(a_1 + \rho)$ and putting (4.12), (4.14), and (4.16) into (4.10) and multiplying by $D(0)D(\rho)(a_2 + \rho)$ gives the following system of two cubic equations in u_1 and u_2:

$$\begin{aligned} 2D(0)D(\rho)(a_1 + \rho)u_1 = {} & c_1b_1D(0)D(\rho) + [(a_2 + \rho)c_1(e_1 - b_1) \\ & - (a_1 + \rho)c_1e_2][e_1(u_1 - u_2) - b_1u_1][a_2b_1u_1 + a_1b_2u_2 \\ & + (a_2 + \rho)c_1b_1[e_1(u_1 - u_2) - b_1u_1]D(0) \\ & + c_1(e_1 - b_1)(a_2b_1u_1 + a_1b_2u_2)D(\rho) \end{aligned} \tag{4.17}$$

$$\begin{aligned} 2D(0)D(\rho)(a_2 + \rho)u_2 &= c_2b_2D(0)D(\rho) + [(a_1 + \rho)c_2(e_2 - b_2) \\ &\quad - (a_2 + \rho)c_2e_1][e_2(u_2 - u_1) - b_2u_2][a_2b_1u_1 + a_1b_2u_2] \\ &\quad + (a_1 + \rho)c_2b_2[e_2(u_2 - u_1) - b_2u_2]D(0) \\ &\quad + c_2(e_2 - b_2)(a_2b_1u_1 + a_1b_2u_2)D(\rho) \end{aligned} \tag{4.18}$$

Multiplying these out and collecting terms, we can rewrite these equations in the following form:

$$\alpha_1u_1^3 + \beta_1u_1^2u_2 + \delta_1u_1u_2^2 + \xi_1u_1^2 + \theta_1u_1u_2 + \Psi_1u_2^2 + \varepsilon_1 = 0 \tag{4.19}$$

$$\alpha_2u_2^3 + \beta_2u_1u_2^2 + \delta_2u_1^2u_2 + \xi_2u_2^2 + \theta_2u_1u_2 + \Psi_2u_1^2 + \varepsilon_2 = 0 \tag{4.20}$$

whose coefficients α_i, β_i, δ_i, ξ_i, θ_i, Ψ_i, and ε_i for i = 1, 2 can be expressed as lengthy expressions in the original parameters.

To solve these equations we start with an initial guess for u_1 and solve the resulting quadratic equation (4.19) in u_2. Hopefully, there is only one sensible value of u_2. We use this value to convert (4.20) into a quadratic equation in u_2 and solve for u_1. With this value of u_1 we go back to (4.19). If this iteration does not converge in a reasonable number of steps, we modify the initial guess of u_1.

§5. NUMERICAL SOLUTION OF THE FINITE COMPETITIVE MODEL

From Section 3, the necessary and sufficient conditions for optimality provide two system equations each with a specified initial condition (2.1), four adjoint equations (3.3)-

(3.6) with their four terminal conditions (3.7), and the two equations for the control variables (3.8) and (3.9).

The algorithm is initiated by guessing values for each competitor's controls for each discrete instant of time during the planning horizon. The system equations are then functions of the state variables only and can be solved numerically forward in time by using the specified initial conditions. At each forward step of the solution, the values of $x_i(t)$ are calculated and stored. The values of the performance indices are calculated during each forward pass.

At terminal time, the terminal conditions on the adjoint variables are calculated and are used in the solution of the adjoint equations backward in time. At each successive time decrement of the backward pass, the appropriate values of the adjoint variables are determined along with a new set of control variable values. When back at the initial time, there is a revised time sequence of values for the control function. These are used in the next forward pass, the objective of which is to evaluate the effect of these new controls on the performance indices. At the end of each forward pass a stopping criterion is evaluated. When the iterative procedure has been terminated, the solution is checked against the Nash equilibrium property. During extensive testing, the algorithm exhibited stability and realistic convergence properties while consistently obtaining Nash solutions.

Using the numerical algorithm, we have investigated many competitive encounters by varying the parameter values individually and in combination. The computer runs provide information on the values of the performance indices, the state and control variables and the adjoint variables at each instant of time during the planning horizon. Computer generated graphs of this information are also obtained.

Limited space does not allow for an extensive presentation of the numerical results. For illustrative purposes, we present two extreme cases.

The first case is a sequence of four encounters where the two competitors are identical except for one attribute, the speed of sales decay. The firms have the same moderate sales response, relatively low competitive reaction and low initial market saturation. Both firms are attempting to maximize profit over the finite planning period. The sales decay parameter for Firm 1 remains at the same moderate level during the four encounters. Firm 2 has a very low rate of decay in encounter 1. This decay increases until it is identical to Firm 1's decay in encounter 3 and is larger than Firm 1's in encounter 4. The parameter values are found in Table 1.

TABLE 1. Encounters 1-4

i	a_i	b_i	e_i	c_i	w_i	J_i	$x_i(0)$	$x_i(T)$
1	.25	.005	.01	200	0	69.55	.10	.020
2	.05	.005	.01	200	0	211.18	.10	.149
1	.25	.005	.01	200	0	76.45	.10	.022
2	.10	.005	.01	200	0	162.45	.10	.091
1	.25	.005	.01	200	0	86.32	.10	.025
2	.25	.005	.01	200	0	86.32	.10	.025
1	.25	.005	.01	200	0	87.90	.10	.025
2	.30	.005	.01	200	0	72.93	.10	.017

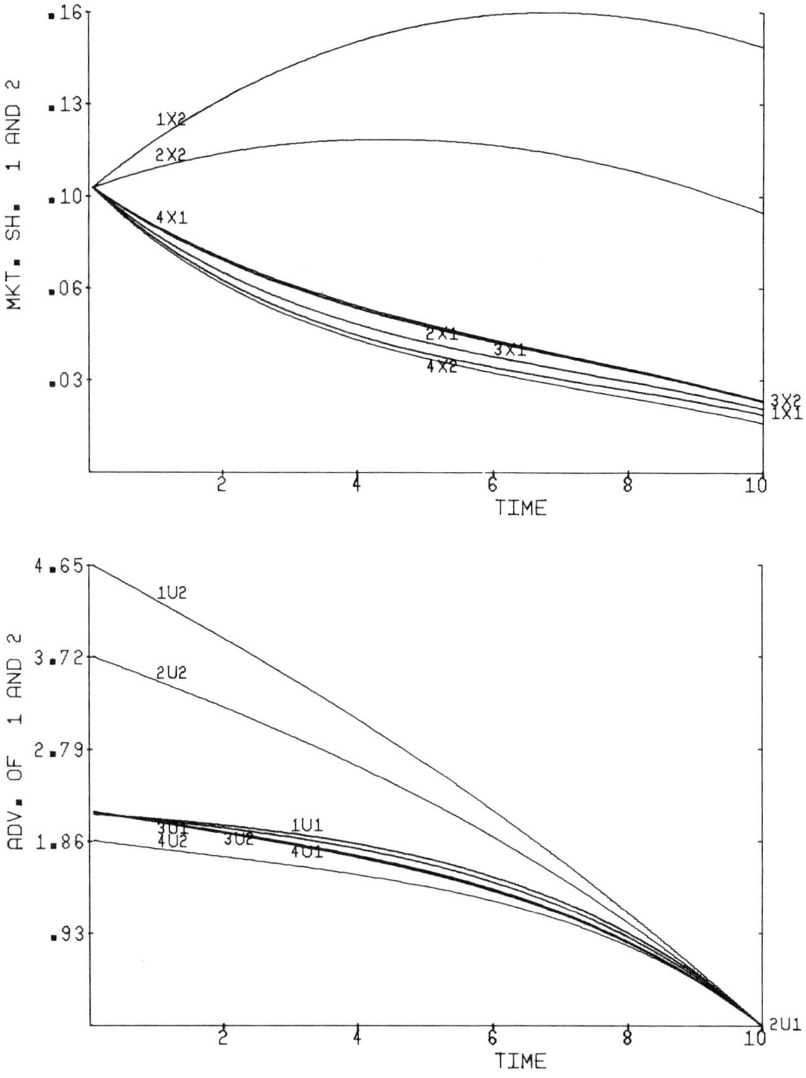

FIGURE 1

The graphical results are presented in Figure 1. The labelling of the curves is iUj for the advertising curves and iXj for the market share curves where i refers to the encounter number, i = 1, ..., 4, and j indicates the firm number, j = 1, 2. For example, in Figure 1a the optimal advertising expenditures path for Firm 2 during encounter 1 is indicated by 1U2. Firm 2's resulting market shares path is labelled 1X2 in Figure 1b.

The time-phased advertising and market share paths should appeal to a realistic explanation from a "marketing common sense" standpoint. Since the firms are concerned with maximizing profit over the current planning period only, it would be expected that the firms should spend their advertising dollars to gain maximum performance index value before the end of the period. To these firms the carryover of market share response to the next period is wasted money. Consequently, advertising expenditures should decrease to zero by the end of the term. In encounter 1, Firm 2 has a much slower decay of market share (greater carryover) than does Firm 1. Although the immediate response to advertising expenditures is identical, Firm 2 obtains significantly greater total effect for each dollar spent. Firm 1 is overwhelmed by Firm 2. Firm 2 is able to obtain a substantially larger total profit and increase its market share by spending more than twice Firm 1's amount initially and tapering to zero by terminal time. Firm 1 finds that a subsistence level of advertising allows it to "maximize" profit in a decidedly dominated situation.

As shown in Table 1, the decay parameter of Firm 1 increases until it becomes larger than 2's decay in encounter 4. Naturally, the advertising and sales paths are quite different in encounters 1 and 4. The profit maximization objective has been investigated extensively by holding all but one parameter fixed and varying the values of that key parameter over a large range. Also, an experimental design was used to perform model testing.

TABLE 2. Encounters 1-4

i	a_i	b_i	e_i	c_i	w_i	J_i	$x_i(0)$	$x_i(T)$
1	.25	.005	.01	200	2550	-72.40	.10	.039
2	.05	.005	.01	200	2550	749.89	.10	.389
1	.25	.005	.01	200	2550	60.65	.10	.062
2	.10	.005	.01	200	2550	494.13	.10	.240
1	.25	.005	.01	200	2550	188.24	.10	.084
2	.25	.005	.01	200	2550	188.24	.10	.084
1	.25	.005	.01	200	2550	202.28	.10	.086
2	.30	.005	.01	200	2550	147.90	.10	.065

The parameter values used in the second set of encounters are presented in Table 2. The corresponding graphs are in Figure 2. These graphs illustrate advertising strategies and market results which are practically opposite to those of the first set of encounters. Of course, the key to this difference is the firms' reorientation to placing heavy importance on terminal market share, as is indicated by the value of the weight, w, in the table.

To illustrate the effect of the competitive interaction term, we present one last set of encounters in Table 3 and Figure 3. As shown in the "e" column of Table 3, Firm 1 is dominant in encounter 1, is identical to Firm 2 in encounter 2, and becomes dominated by Firm 2 in encounters 3 and 4.

Over these encounters, Firm 2 changes position from one in which it cannot maintain its initial market share during encounter 1 to the position where it has a 70% market share compared to Firm 1's 4% market share in encounter 4.

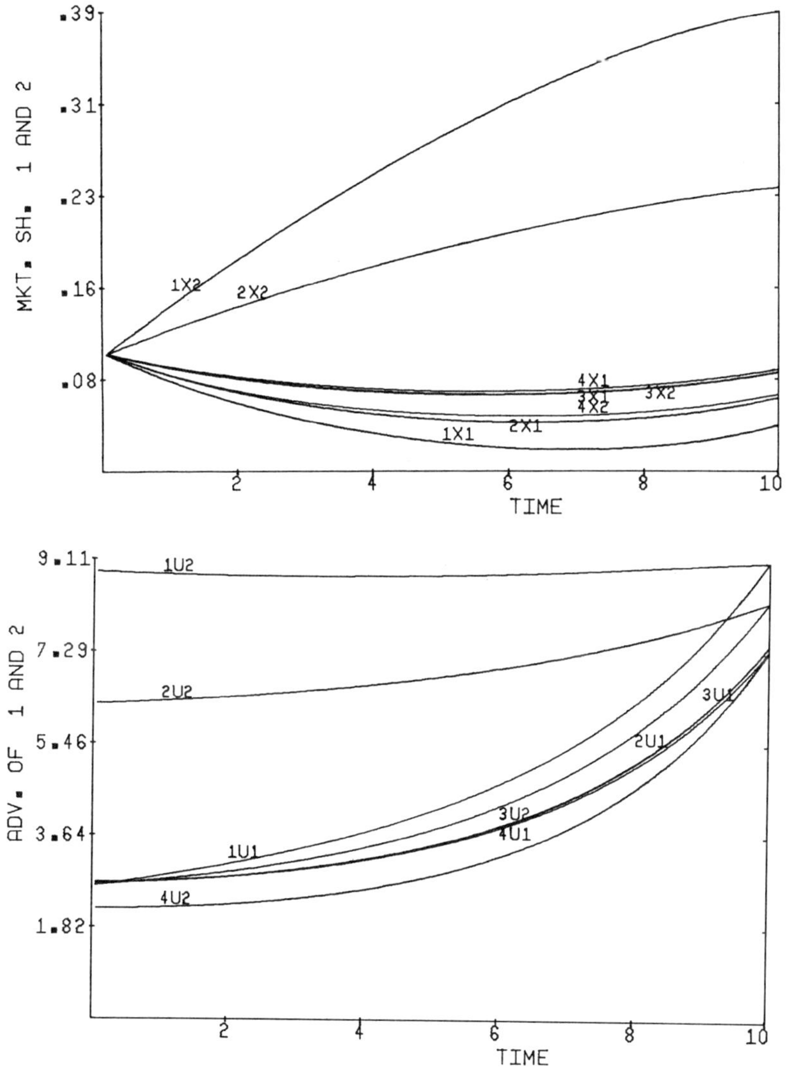
.39
.31
.23
.16
.08
MKT. SH. 1 AND 2
1X2
2X2
4X1
3X1
4X2
3X2
1X1
2X1
2
4
6
8
10
TIME
9.11
7.29
5.46
3.64
1.82
ADV. OF 1 AND 2
1U2
2U2
3U1
2U1
3U2
4U1
1U1
4U2
2
4
6
8
10
TIME

FIGURE 2

TABLE 3. Encounters 1-4

i	a_i	b_i	e_i	c_i	w_i	J_i	$x_i(0)$	$x_i(T)$
1	.10	.005	.010	200	2550	703.28	.30	.318
2	.10	.005	.005	200	2500	791.24	.30	.167
1	.10	.005	.010	200	2500	407.91	.30	.244
2	.10	.005	.010	200	2500	407.91	.30	.244
1	.10	.005	.010	200	2500	-53.82	.30	.165
2	.10	.005	.015	200	2500	298.22	.30	.396
1	.10	.005	.010	200	2500	-997.35	.30	.039
2	.10	.005	.020	200	2500	-62.87	.30	.702

Comparison of the advertising strategies resulting from profit maximization (Figure 1) and those resulting from having a heavy emphasis on terminal market share (Figure 2) indicates the tremendous differences between advertising strategies under these two performance criteria. Naturally, less extreme weighting of the components of the performance indices would produce results which are more realistic for most actual marketing organizations.

Many advertising executives maintain the need to have a very strong impact in a communication media or to stay out of that media except for some bare subsistence level spending. The benefit of dominating an industry through advertising has been illustrated strongly by the last set of encounters which was based on variation of the reaction terms. The effect of the reaction terms will vary depending upon the values of the other parameters and the performance index weighting. However, the benefit of advertising dominance will exist in every case.

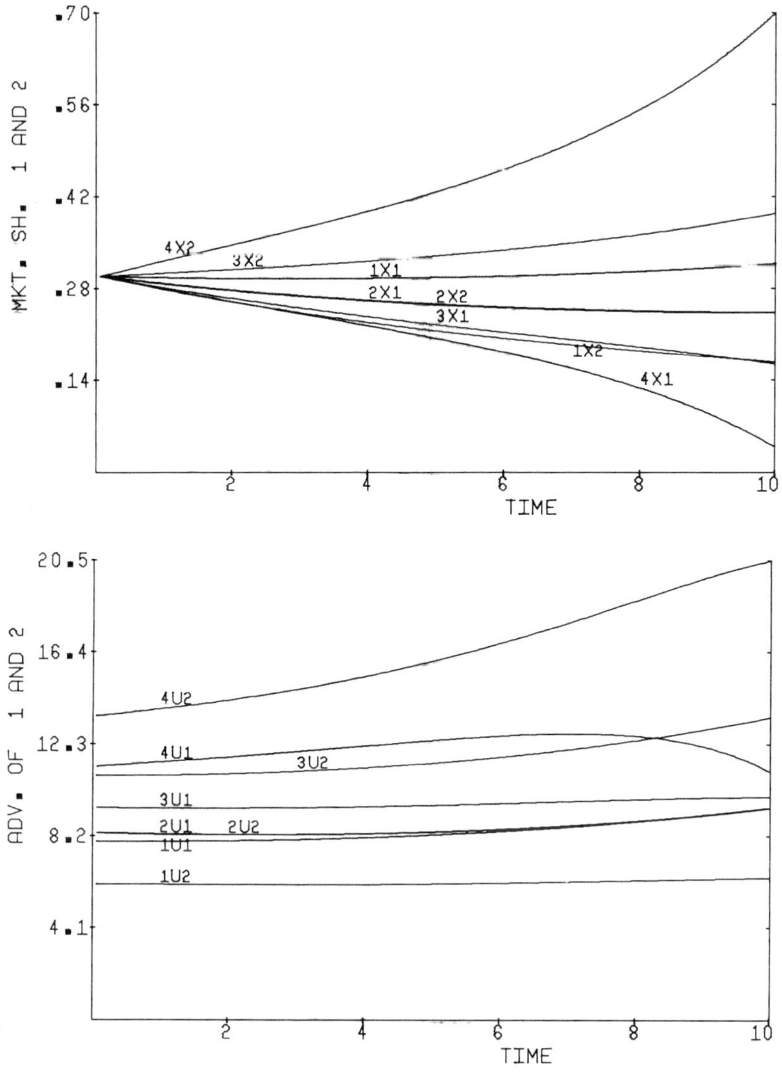

FIGURE 3

§6. CONCLUSIONS

The model presented in this paper has the conceptual advantage of combining logically several critical dimensions of advertising competition. A benefit of this model lies in its ability to represent the marketing characteristics of market share response to advertising, decay of market share due to forgetting, the saturation effect, and the effect of sales switching due to advertising dominance within the most natural environment of advertising--dynamic competition with multiple objective performance indices.

We are encouraged by the numerical results to the finite horizon problem. Our next stage in the investigation will be to solve numerically the infinite horizon problem as proposed in Section 4 and to compare the results to the solution of the original problem when a greatly lengthened, but finite, time period is used. We are also looking forward to the research of a particular industrial situation which coincides closely with the assumptions of this model.

ACKNOWLEDGMENTS

The first two authors are indebted to Grants A9270 and A4619 from the National Research Council of Canada for partial support of this research. The third author's research was supported by the Office of Naval Research through a contract with Carnegie-Mellon University.

REFERENCES

1. Vidale, M. L., and Wolfe, H. B., An operations research study of sales response to advertising, Operations Res., 5, 370-381 (1957).
2. Sethi, S. P., Optimal control of the Vidale-Wolfe advertising model, Operations Res., Vol. 21, 998-1013 (1973).
3. Sethi, S. P., Some explanatory remarks on the optimal control of the Vidale-Wolfe advertising model, Operations Res., 22, 1119-1120 (1974).

4. Sethi, S. P., Dynamic optimal control models in advertising: A survey, SIAM Review, Vol. 19, No. 4, 685-722 (October 1977).

5. Deal, K. R., and Zionts, S., A differential games solution to the problem of determining the optimal timing of advertising expenditures, Proc. Second Annual Northeast Regional AIDS Conference (Kingston, R.I., 1973), American Institute of Decision Sciences, Atlanta, Ga., 1973.

6. Deal, K. R., A differential games solution to the problem of determining the optimal timing of advertising expenditures, unpublished Ph.D. dissertation, State University of New York at Buffalo, Buffalo, 1975.

7. Starr, A. W., Non-zero-sums differential game: Concepts and models, Harvard University, Division of Engineering and Applied Physics, Technical Report No. TR590, 1969.

8. Case, J., Economics and the Competitive Process, New York University Press, 1978.

ADVERTISING GAMES IN DUOPOLY MARKET EXPANSION

C. Wrather and P. L. Yu

Graduate School of Business
University of Texas, Austin, Texas

and

School of Business, University of Kansas
Lawrence, Kansas

ABSTRACT

A company which produces a particular product can emphasize two different features in its advertising. It can emphasize the *product features* of its product in an attempt to enlarge the size of the product market, or it can emphasize the *brand features* in an attempt to increase market share at the expense of its competitors. Thus, market expansion games naturally embody the characteristics of partial cooperation and partial competition. In this article two models are suggested to study such games. Optimal response strategies are derived, equilibrium points are discussed, and the stability of the strategies are considered. Incentives for cooperation, conspiracy, or merger are also discussed. This is a first step in using game theoretical models in the context of advertising interactions where partial cooperation and partial competition are involved. Various research problems are outlined at the conclusion.

§1. INTRODUCTION

Consider an advertising game between two competing companies who produce different brands of a similar product. One of the important strategic advertising decisions for a firm is to determine what groups of people the advertising effort should be aimed at. As described in Figure 1, the future users of our particular brand can come from three groups: those who are currently nonusers of the product, those who are currently users of the product but who do not use our brand, and those who currently use our brand of the product.

The object of advertising is to increase the number of individuals who flow into the "users of our brand" category. This can be done with two kinds of strategies. First, a firm can use a *product strategy* by attempting to increase the number of new users of the product and keep the current product users in the market. In Figure 1, a product strategy attempts to increase the net flow of individuals to the right in channel 1. The philosophy behind such a strategy is the belief that by increasing the number of individuals using the product, some of these new users of the product will try our brand.

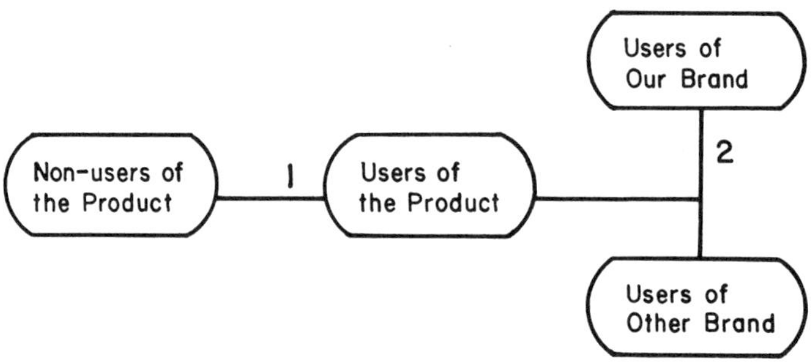

FIGURE 1

Familiar examples of product strategy advertisement are the "industry" advertising done by milk producers' associations and the Florida orange juice producers.

A different strategy is a *brand strategy* which attempts to persuade the current nonusers of our brand (who are the product users) to our brand next, and to keep our brand users from switching to the other brand next. In terms of Figure 1, a brand strategy is aimed at increasing the net flow of individuals upward along channel 2. Aaker and Myers [1] refer to reducing the downward flow along channel 2 as a "brand loyalty strategy" or a "defensive strategy," and increasing the upward flow of individuals as a "new trier strategy." We shall refer to these strategies collectively as brand strategies.

Observe that using a product strategy benefits both companies and using a brand strategy benefits one at the other's expense. In this report we suggest two new game theoretical models (Section 2) for this partially competitive (brand strategy) and partially cooperative (product strategy) phenomenon to answer the following important questions.

(i) With respect to its opponent's strategy, what is the optimal response strategy for each company? Is there any equilibrium pair of optimal response strategies? In Section 3 explicit optimal response strategies will be derived, which lead to discussion of Nash equilibrium pairs of strategies. Conditions for Nash equilibrium pair to exist and/or to be unique are derived for both pure and mixed strategies sense.

(ii) When each company interactively and iteratively maximizes its sales while taking its opponent's strategy into consideration, is there a convergent or stable equilibrium pair of strategies? In Section 4, "stable" and "absolutely stable" pairs of strategies are defined. It is shown that a stable pair of strategies is, under a mild condition, a Nash equilibrium pair. But a Nash equilibrium pair may not be stable. Conditions for a game to have stable and/or absolutely stable pair of strategies are derived.

(iii) Instead of competition, can the companies cooperate, conspire or merge so that both attain larger sales? This will be discussed in Section 5.

At the end of this report, we shall mention some possible but important extensions of our models for further studies (Section 6).

Before we close this section, it is important to note that a game theoretical model has been developed by Friedman [3] for describing two competing firms' allocations of advertisement budgets over a set of sales districts. Each sales district has a *fixed* number of purchasers and the brand share in each district is divided between the firms proportional to the advertising expenditures of each firm. An implicit assumption of this formulation is that prior to advertising, sales are divided equally between the firms. More recently, Shane [11] has elegantly generalized Friedman's model so that the assumption of equal market share prior to advertising can be dropped. Besides, Shane also has developed a dynamic allocation model. However, Shane still assumes that a fixed number of purchasers are divided between the two competitors as Friedman does. Thus, these models only work for totally competitive (noncooperative constant sum game) situations. The present proposed model, however, includes in its formulation the dual role of advertising--to increase market share (brand strategy) and to increase the size of the market (product strategy). Under this formulation an acceptable situation for a dominant firm might be to concentrate on expanding the market for the product in preference to counteracting a smaller competitor's modest attempts at stealing away its customers. For this reason we are convinced that the current formulation can more fully describe real world competitive behavior in advertising. To our best knowledge, the models to be presented are new. The important problems which we address, such as optimal response strategies, equilibrium, stability, cooperation or conspiracy, are not issues in Shane's paper.

Clemhout, Leitmann, and Wan [14], Leitmann and Schmitendorf [15] have worked out some differential games models for advertisement. Their main concerns are to obtain Nash equilibrium solutions for their competitive models. Clearly, our emphasis is different from theirs. Note that our discussion on solution concepts can be extended to dynamic cases by considering the criteria space (the set of all possible outcomes of the games).

§2. TWO GAME MODELS

Let D_1 and D_2 be the total advertising budgets for Company 1 and Company 2, respectively. Let $x = (x_1, x_2)$ and $y = (y_1, y_2)$ be the advertising strategies in terms of dollar expenditure of Company 1 and Company 2, respectively, with x_1

and y_1 representing brand strategies and x_2 and y_2 for product strategies. Thus

$$x_1 + x_2 \leqq D_1 \qquad \text{and} \qquad y_1 + y_2 \leqq D_2$$

Observe that each customer not only decides his or her own next brand choice, but also influences his or her relatives and friends' choice. In order to capture this behavioral effect, let T_{ij} be the *transition propensity of sales expansion* from brand i to brand j. That is, given one unit of current sales of brand i, T_{ij} gives the potential sales of brand j for the next period.[1] Now let $s = (s_1, s_2)$ be the current sales, with s_i being the ith company's sales. Then the sales for the next period $\hat{s} = (\hat{s}_1, \hat{s}_2)$ is given by

$$\hat{s} = s \cdot T = (\sum_i s_i T_{i1}, \sum_i s_i T_{i2}) \tag{1}$$

where T is the matrix with its component represented by T_{ij}.

Note that the above quantiative description of consumers' behavior can be justified by a multiple type branching process model of market growth (see Wrather [13]) which contains the well-known Markov chain brand switch model of Maffei [7] as a special case.

---

[1]Alternatively, one can let s_{ij} be a random variable representing the number of units brand j purchased next period by: (1) those individuals who purchased brand i in the current period and (2) those new purchasers of brand j next period who were influenced by individuals who purchased brand i in the current period. Then define $T_{ij} = E[s_{ij}]/s_i$ which gives a measure of the propensity for a brand i purchase in current period to yield a brand j purchase next period.

Clearly, T_{ij} is a function of the advertising strategy (x, y). In order to assess the functional relationship between T_{ij} and (x, y), recall that (see Figure 1) a product strategy benefits the entire industry (both companies) while a brand strategy benefits one at the other's expense. Thus, we can express a general functional relation as follows:

$$T_{ij}(x,y) = \begin{cases} (1 + f(x_2,y_2))m_{ij} + g(x_1,y_1) & \text{if } j = 1 \\ (1 + f(x_2,y_2))m_{ij} - g(x_1,y_1) & \text{if } j = 2 \end{cases} \quad (2)$$

where $i,j \in \{1,2\}$ and m_{ij} are constants.

Observe that, if $f = g = 0$ when $x = y = 0$, then m_{ij} may be interpreted as the *status-quo transition propensity of sales expansion from brand* i *to* j. That is, in the absence of advertising, existing market forces will generate potential sales of m_{ij} units of brand j in the next period for each unit of current sales of brand i.

Also observe that $f(x_2, y_2)$ is the cooperative *sales expansion factor* which benefits both companies through the product strategy x_2 and y_2; while $g(x_1, y_1)$ is the competitive *sales factor* which measures the *relative sales gain* for Company 1. Note that g is a function of brand strategy x_1 and y_1. If $g(x_1, y_1) > 0$ (or < 0) then the competitive factor favors (or disfavors) Company 1.

Combining (1) and (2), one obtains the sales of the next period $\hat{s} = (\hat{s}_1, \hat{s}_2)$ as

$$\hat{s}_1(x,y) = (1 + f(x_2,y_2))q_1 + Sg(x_1,y_1) \quad (3)$$

where $q_1 = s_1m_{11} + s_2m_{21}$ (which may be interpreted as the next period status-quo sales for Company 1 in the absence of any further advertising), and $S = s_1 + s_2$ is the current total sales of the two companies (i.e., the current market size); and

$$\hat{s}_2(x,y) = (1 + f(x_2,y_2))q_2 - Sg(x_1,y_1) \tag{4}$$

where $q_2 = s_1m_{12} + s_2m_{22}$ (the next period status-quo sales for Company 2).

Observe that q_1 and q_2 are positive; otherwise (3) and (4) do not make sense.

Note that $\hat{s}$ depends not only on f (cooperation factor) and g (competing factor), but also on the existing status quo market position q_1 and q_2.

The following assumptions will be justified shortly.

Basic Assumptions

(i) $f(x_2, y_2)$ is a strictly increasing function of $x_2 + y_2$ for $x_2 + y_2 \in [0, D_1 + D_2]$.

(ii) Company 1 wants to maximize $\hat{s}_1$ and Company 2 wants to maximize $\hat{s}_2$.

Note that (i) essentially says that market potential is large enough not to be saturated by advertising activities within the maximum budget limit. Thus, as (cooperative) product advertising expenditures $x_2 + y_2$ are increased, the product sales $\hat{s}_1 + \hat{s}_2 = (1 + f_2(x_2,y_2))(q_1 + q_2)$ also increase. Assumption (i) is not unreasonable since the demand for almost any product can be created or promoted.

Assumption (ii) will greatly simplify our analysis, although the maximization of sales is clearly open to criticism. Arguments can be made supporting the maximization of profits or even market share. We assume, however, that in the long run profits are an increasing function of sales, thus, maximization of profits and of sales yields the same strategies. The assumption allows us to avoid complicating our presentation with revenue/cost analysis.

For convenience, we introduce

Definition 2.1 The advertising game model specified by (3) and (4) and the basic assumptions will be called *Model 1*.

In order to obtain analytical results in closed form, we shall further assume that f and g are "quadratic" in the following sense:

$$f(x_2,y_2) = d(x_2 + y_2) - e(x_2 + y_2)^2 \tag{5}$$

where d and e are nonnegative constants. (Note: In order to satisfy Basic Assumption (i), f attains its maximum at $x_2 + y_2 = d/2e \geq D_1 + D_2$.) Also,

$$g(x_1,y_1) = (a_1x_1 - b_1x_1^2) - (a_2y_1 - b_2y_1^2) + \begin{cases} c_1(x_1 - y_1)^2, & \text{if } x_1 \geq y_1 \\ -c_2(x_1 - y_1)^2 & \text{if } x_1 < y_1 \end{cases} \tag{6}$$

where a_i, b_i, c_i, i = 1, 2 are nonnegative constants with $c_i \leq b_i$; and unless otherwise specified, c_i are assumed strictly positive so that the interaction effect between x_1 and y_1 can be studied.

Observe that f in (5) is concave increasing within the budget limit reflecting diminishing marginal sales to advertising effort. This property is supported by Simon [12].

The competitive factor $g(x_1, y_1)$ not only enjoys the diminishing marginal sales to advertising effort but also incorporates a "competitive market position" concept. That is, the firm spending the most on a brand strategy, thus enjoying a position as the "competitive leader" in the eyes of the public, benefits from this by an increase in the effectiveness of its brand strategy.

Observe that all constants in (5) and (6) are determined by existing market positions and advertising efficiency. We treat them as parameters so that a large class of problems can be studied. In the case where f and g are not quadratic as in (5) and (6), we may regard (5) and (6) as a first order or second order approximation of the real relationship. For convenience, we introduce

Definition 2.2 The advertising game model specified by (3)-(6) and the basic assumptions will be called *Model 2*.

In order to facilitate presenting the results of our analysis we shall start with Model 2 which is more specific, then generalizing the results to Model 1.

§3. OPTIMAL RESPONSE STRATEGIES AND EQUILIBRIUM POINTS

In this section, we shall study optimal strategies for responding to the opponent's advertising strategies. Recall that each company's sales depends on both companies' advertising strategies, thus one company's strategy must be chosen with its competitor's strategy in mind. The interesting question arises as to the existence of equilibrium points (in the Nash sense) and the ability to construct them. By an equilibrium point we mean a pair of strategies (x^*, y^*) such that x^* is optimal for Company 1 when Company 2 uses y^*, and y^* is optimal for Company 2 when Company 1 uses x^*. (See Definition 3.1.)

One of the important consequences of Assumption (i) is:

<u>Theorem 3.1</u> If (x_1^*, x_2^*) and (y_1^*, y_2^*) are optimal strategies (in terms of maximizing $\hat{s}_1$ and $\hat{s}_2$, respectively), then $x_1^* + x_2^* = D_1$ and $y_1^* + y_2^* = D_2$.

<u>Proof</u> Suppose that $x_1^* + x_2^* < D_1$. Let $x_2^0 = D_1 - x_1^* > x_2^*$. By substituting $x^0 = (x_1^*, x_2^0)$ into (3) and invoking (increasing) assumption (i), we see that $\hat{s}_1(x^0, y) > \hat{s}_1(x^*, y)$ for all possible y, contradicting the optimality of x^*. The remaining part of the theorem can be proved similarly. Q.E.D.

Theorem 1 greatly simplifies our analysis, because in searching for optimal strategies we can, without loss of generality, assume that $x_1 + x_2 = D_1$ and $y_1 + y_2 = D_2$.

Thus, hereafter we shall represent $x_2 = D_1 - x_1$ *and* $y_2 = D_2 - y_1$; *and to simplify the notation, we shall replace* x_1 *and* y_1 *by* x *and* y, *respectively.* Thus (x, y) will hereafter represent the brand strategy of Company 1 and 2, respectively.

Proceeding with Model 2, for the sake of smooth presentation, we shall further assume that f is linear, that is, $e = 0$ in (5). The results for $e \neq 0$ will be given in Appendix B.

By invoking standard optimality conditions, one obtains in closed form the optimal response strategies for various parameters as follows. (The details of their derivation are supplied in Appendix A. Their interesting interpretation will be given shortly.)

<u>Theorem 3.2</u> Let $w_1 = a_1S - dq_1$. The optimal response strategy $x^*(y)$ for Company 1 for each competing strategy y of Company 2, depending on the parameters, is given by:

(i) if $w_1 \leqq 0$, then

$$x^*(y) = \begin{cases} 0, & \text{if } 0 \leqq y \leqq -w_1/2Sc_2 \\ \min\{\underline{x}^*(y), D_1\}, & \text{if } y \geqq -w_1/2Sc_2 \end{cases} \tag{7}$$

$$\text{where } \underline{x}^*(y) = \frac{c_2}{b_1 + c_2} y + \frac{w_1}{2S(b_1 + c_2)} \tag{8}$$

(ii) if $0 < w_1 < 2Sb_1D_1$, then

$$x^*(y) = \begin{cases} \min\{\bar{x}^*(y), D_1\}, & \text{if } 0 \leqq y \leqq w_1/2Sb_1 \quad (9) \\ \min\{\underline{x}^*(y), D_1\}, & \text{if } y > w_1/2Sb_1 \quad (10) \end{cases}$$

$$\text{where } \bar{x}^*(y) = \frac{-c_1}{b_1 - c_1} y + \frac{w_1}{2S(b_1 - c_1)} \tag{11}$$

and $\underline{x}^*(y)$ is given by (8).

(iii) if $w_1 \geqq 2Sb_1D_1$, then $x^*(y) = D_1$ for all $y \in [0, D_2]$

Observe that the optimal response strategy depends not only on y but also on the parameters (a_i, b_i, c_i, d, e, s_i, etc.) which describe the market positions and advertising efficiency.

Theorem 3.2 enjoys an interesting and intuitively appealing interpretation. Before we discuss it, we shall make (7),

(9), and (10) clearer. Note that (7) is a short-cut of the following expression

$$x^*(y) = \begin{cases} \underline{x}^*(y) & \text{if } \dfrac{-w_1}{2Sc_2} \leqq y \leqq \dfrac{-w_1}{2Sc_2} + \dfrac{D_1(b_1 + c_2)}{c_2} \\ D_1 & \text{if } y \geq \dfrac{-w_1}{2Sc_2} + \dfrac{D_1(b_1 + c_2)}{c_2} \end{cases}$$

because $\underline{x}^*(y) \leqq D_1$ iff

$$y \leqq \frac{-w_1}{2Sc_2} + \frac{D_1(b_1 + c_2)}{c_2}$$

Similarly (9) and (10) can be rewritten. We shall not stop to do so.

Now let us consider the following interpretation of the variable $w_1 = a_1S - dq_1$. Note from (3) and (6) that for Company 1, a_1 is a first order approximation of the marginal increase in sales in this period per dollar of expenditure on a brand strategy, per sale, in the last period. Thus a_1S represents an approximation of the marginal sales gain per dollar of brand advertising. Similarly, from (3) and (5), d is a first order approximation of the marginal sales gain this period per dollar of expenditure on a product strategy per status-quo sale. Thus, q_1d is an approximation to the marginal sales gain per dollar expenditure on a product strategy. Hence, w_1 is a measure of the marginal advantage of using a brand strategy as opposed to a product strategy.

Thus, in (i) of Theorem 3.2, when $w_1 \leqq 0$ (that is, the relative marginal advantage of using a brand strategy is negative), Company 1 should allocate nothing to a brand strategy (and everything to a product strategy) until its competitor spends more than $-w_1/2Sc_2$ on a brand strategy. In the latter case, Company 1 should increase its brand strategy expenditures proportionally with those of its competitor.

In (iii), when $w_1 \geqq 2Sb_1D_1 > 0$ (that is, the relative marginal advantage of using a brand strategy is sufficiently large), Company 1 should devote its entire budget to a brand strategy.

In order to understand (ii), observe that $x^*(y) = \bar{x}^*(y) = \underline{x}^*(y) = y$ when $y = w_1/2Sb_1$. Thus a typical $x^*(y)$ can be depicted as in Fig. 2. Thus when (ii) of Theorem 3.2 prevails (that is, the relative marginal advantage of using a brand strategy is positive, but not exceeding $2Sb_1D_1$), then Company 1 should devote its entire budget or $\bar{x}^*$ (whichever is smaller) to a brand strategy as long as Company 1 is the competitive leader (i.e., $x^* > y$). Note that $\bar{x}^*$ is decreasing in y. That is, as long as the Company is the competitive leader, it should concentrate more on developing the product market than worrying about

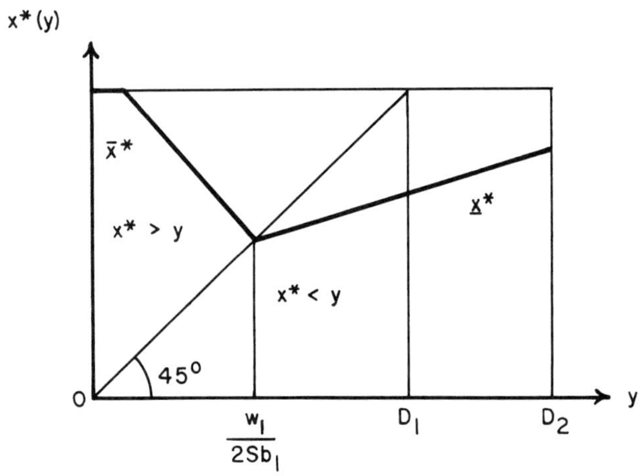

FIGURE 2

increasing market share. Note that when y is increased, Company 2's product effort $(D_2 - y)$ is decreased. It is to Company 1's advantage to increase product strategy expenditures (thus decrease x). However, if Company 1 has lost its competitive leadership (i.e., $y > w_1/2Sb_1$), it should engage in direct brand competition with its competitor by allocating $\underline{x}^*$ or D_1 to a brand strategy, whichever is smaller. Note that $\underline{x}^*$ is increasing in y. Thus, if Company 2 increases its brand strategy expenditures, Company 1 should also increase its brand effort.

Because of the symmetry of the problem the optimal response strategy $y^*(x)$ for Company 2 is symmetric to $x^*(y)$ in Theorem 3.2 (by interchanging subindex 1 and 2) and enjoys the same interpretation. We have:

<u>Theorem 3.3</u> Let $w_2 = a_2S - dq_2$.

(i) If $w_2 \leqq 0$, then

$$y^*(x) = \begin{cases} 0 & \text{if } 0 \leqq x \leqq -w_1/2Sc_1 \\ \min\{\bar{y}^*(x), D_2\} & \text{if } x \geqq -w_2/2Sc_1 \end{cases}$$

where

$$\bar{y}^*(x) = \frac{c_1 x}{b_2 + c_1} + \frac{w_2}{2S(b_2 + c_1)}$$

(ii) If $0 < w_2 < 2Sb_2D_2$, then

$$y^*(x) = \begin{cases} \min\{\underline{y}^*(x), D_2\} & \text{if } 0 \leqq x \leqq w_2/2Sb_2 \\ \min\{\bar{y}^*(x), D_2\} & \text{if } x > w_2/2Sb_2 \end{cases}$$

where

$$\underline{y}^*(x) = \frac{-c_2}{b_2 - c_2} x + \frac{w_2}{2S(b_2 - c_2)}$$

(iii) If $w_2 > 2Sb_2D_2$, then $y^*(x) = D_2$ for all $x \in [0,D_1]$

Optimal response strategies in the more general setting--Model 1--exist whenever we assume that both f and g are continuous. However, in the more general case, the nice expressions of Theorems 3.2 and 3.3 disappear, and locating optimal response strategies constitutes a mathematical programming problem; we shall not elaborate on it here.

We now turn our attention to equilibrium points.

Definition 3.1 (x^0, y^0) is a pair of *equilibrium strategies* if and only if the following hold:

$\hat{s}_1(x^0,y^0) \geqq \hat{s}_1(x,y^0)$ for all $x \in [0,D_1]$

$\hat{s}_2(x^0,y^0) \geqq \hat{s}_2(x^0,y)$ for all $y \in [0,D_2]$

Definition 3.1 essentially says that if either company unilaterally deviates from the equilibrium pair that company will be "worse off" in its sales. Again we start with Model 2.

Theorem 3.4 There exists a *unique* pair of equilibrium strategies for every game in the form of Model 2 with $e = 0$.

Proof Our proof is based on Theorems 3.2 and 3.3. To be complete, we need consider all possible combinations of cases (i)-(iii) in both theorems. However, any combination with case (iii) of Theorem 3.2 or 3.3 can be proved easily; and the combinations of case (i) of Theorem 3.2 (or Theorem 3.3) with case (i) or (ii) of Theorem 3.3 (or Theorem 3.2) can be treated similarly to the combination of case (ii) for both Theorems

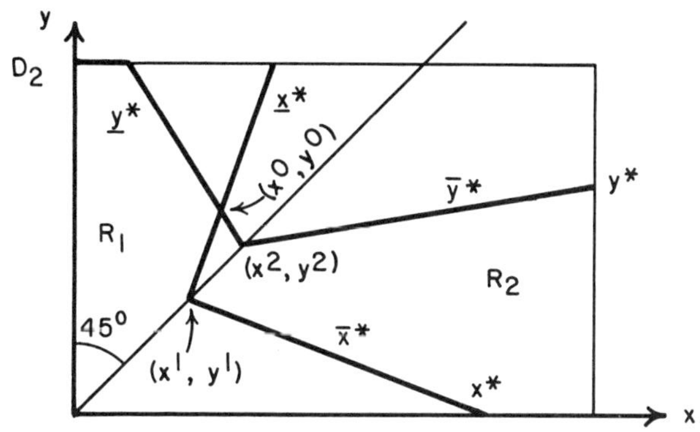

FIGURE 3

3.2 and 3.3. Thus, we shall focus on the last combination and leave the rest to the reader.

Referring to Figure 3, let

$$R_1 = \{(x,y)\,|\,y \geqq x,\ 0 \leqq x \leqq D_1,\ 0 \leqq y \leqq D_2\}$$

and

$$R_2 = \{(x,y)\,|\,y < x,\ 0 \leqq x \leqq D_1,\ 0 \leqq y \leqq D_2\}$$

It can be verified (see Appendix A) that both $x^*(y)$ and $y^*(x)$ are continuous in their respective domains. Furthermore, in R_1,

$$x^*(y) = \min\{\underline{x}^*(y), D_1\} \qquad y^*(x) = \min\{\underline{y}^*(x), D_2\}$$

where $\underline{x}^*(y)$ is strictly increasing and $\underline{y}^*(x)$ is strictly decreasing; and in R_2

$$x^*(y) = \min\{\bar{x}^*(y), D_1\} \qquad y^*(x) = \min\{\bar{y}^*(x), D_2\}$$

where $\bar{x}^*(y)$ is strictly decreasing, and $\bar{y}^*(x)$ is strictly increasing.

First observe that (x^0, y^0) is a pair of equilibrium strategies if and only if $x^0 = x^*(y^0)$ and $y^0 = y^*(x^0)$.

(Otherwise, the inequalities of Definition 3.1 cannot be satisfied.) Thus, our problem reduces to demonstrating the existence of a unique intersection point of the graph of $x^*(y)$ and the graph of $y^*(x)$.

Define $x^* \circ y^* : [0,D_1] \to [0,D_1]$ by $x^* \circ y^*(x) = x^*(y^*(x))$. Since both x^* and y^* are continuous, $x^* \circ y^*$ is continuous. The Brower fixed point theorem immediately implies the existence of an x^0 so that $x^0 = x^* \circ y^*(x^0)$. Define $y^0 = y^*(x^0)$. Thus (x^0, y^0) is a pair of equilibrium strategies. It remains to show that (x^0, y^0) is unique.

Suppose that $(x^0, y^0) \in R_1$ (refer to Figure 3). By monotonicity of $x^*(y)$ and $y^*(x)$ in R_1, the graphs of $x^*(y)$ and $y^*(x)$ cannot intersect at any other point than (x^0, y^0) in R_1. Let the graphs of $x^*(y)$ and $y^*(x)$ intersect the line $x = y$ at (x^1, y^1) and (x^2, y^2), respectively. (Such intersections are assured because we are dealing with the combination of two functions satisfying Case (ii) of Theorems 3.2 and 3.3.) By the monotonicity of $x^*(y)$ and $y^*(x)$, it is readily verified that $y^2 \geqq y^1$. Then, by the monotonicity of x^* and y^* in R_2, the graph of $y^*(x)$ will lie strictly above that of $x^*(y)$ (i.e., the y-coordinate of the graph of $y^*(x)$ is always larger than that of $x^*(y)$). That is, the graph of $x^*(y)$ cannot intersect the graph of $y^*(x)$ in R_2. Thus (x^0, y^0) is the unique intersection point.

Similarly, if $(x^0, y^0) \in R^2$, one can show that (x^0, y^0) is the unique intersection point. We shall not carry out the proof here. Q.E.D.

Note that the pair of equilibrium strategies in Model 2 can easily be obtained by applying Theorems 3.2 and 3.3.

(See Example 5.1.)

We now state results for models more general than Model 2.

Observe that the result of Theorem 3.4 is based on continuity and monotonicity, rather than on the piecewise linearity of $x^*(y)$ and $y^*(x)$. This observation yields

<u>Theorem 3.5</u> Let $x^*(y)$ and $y^*(x)$ be the optimal response strategies of Companies 1 and 2, respectively, in Model 1. Suppose that $x^*(y)$ and $y^*(x)$ are continuous functions in their respective domains and enjoy the following monotonicity properties:

(i) $x^*(y)$ is nondecreasing in y as long as $x^*(y) \leq y$ and is nonincreasing in y as long as $x^*(y) > y$

(ii) $y^*(x)$ is nonincreasing in x as long as $x \leq y^*(x)$ and is nondecreasing in x as long as $x > y^*(x)$

Then there is a unique pair of equilibrium strategies for the advertising game.

Observe that conditions (i) and (ii) correspond to the monotonicity in R_1 and R_2, respectively, specified in the proof of the previous theorem.

Observe that if f is a concave and continuous function in its domain and g is continuous, concave in x and convex in y in its domain, then the payoffs $\hat{s}_1(x, y)$ and $\hat{s}_2(x, y)$ [see (3) and (4)] will both be continuous. Furthermore, when y is fixed, $\hat{s}_1(x, y)$ is concave in x; and when x is fixed, $\hat{s}_2(x, y)$ is concave in y. A result by Nikaido and Isoda [8] shows that under these conditions a pair of equilibrium strategies exists. We summarize this as

Theorem 3.6 In Model 1, there always exists a pair of equilibrium strategies when f is concave and continuous in its domain and g is continuous, concave in x and convex in y in its domain.

Suppose that f and g do not satisfy the conditions specified in Theorem 3.6. Pairs of equilibrium strategies may not exist as defined in Definition 3.1 (i.e., in a "pure" strategy sense). However, if we allow randomization in strategies, we can obtain a more general result.

Definition 3.2 A "mixed" advertising strategy for Company 1 or 2 is a probability distribution ϕ on $[0, D_1]$ or ψ on $[0, D_2]$, respectively. The expected sales for a pair of strategies (ϕ, ψ) are defined by

$$\bar{s}_i(\phi,\psi) = \underset{(\phi,\psi)}{E} \hat{s}_i(x,y) \qquad i = 1, 2$$

(ϕ^0, ψ^0) is a pair of equilibrium mixed strategies if

$$\bar{s}_1(\phi^0,\psi^0) \geqq \bar{s}_1(\phi,\psi^0) \quad \text{and} \quad \bar{s}_2(\phi^0,\psi^0) \geqq \bar{s}_2(\phi^0,\psi)$$

for all possible (ϕ, ψ).

Following the result of Glicksburg [4], we obtain

Theorem 3.7 If both f and g are continuous in their domain, then there always exists a pair of equilibrium mixed strategies.

§4. STABLE EQUILIBRIUM STRATEGIES

In the present marketing game, suppose that each company tries to maximize its sales based on a guess about its competitor's strategy. Then each company attempts to maximize its sales by taking into account the strategies of its competitor, who in turn maximizes its sales by taking into account the strategies of its competitor, who in turn ---, etc. Two serious questions arise. Does this interactive and iterative maximization process converge? Is the limiting solution of this process independent of the initial guesses? In a free enterprise system, the market adjustment mechanism plays a key role for equilibrium. The answer to the above two questions is thus important to our analysis. If the process converges to a solution which is independent of its initial guess, the market share will have a steady-state distribution, and this distribution can be viewed as a measure of the relative market strength of the companies. In this section we shall address the above questions. First, let us make the above concept precise. Throughout this section we shall assume that the optimal response strategies $x^*(y)$ and $y^*(x)$ exist and are functions.

Let $x^1 \in [0, D_1]$ be an initial point. Let $y^1 = y^*(x^1)$. Sequentially we define

$$x^{n+1} = x^* \circ y^*(x^n) \qquad n = 1, 2, \ldots$$

$$y^{n+1} = y^* \circ x^*(y^n) \qquad n = 1, 2, \ldots$$

(Clearly, one could start with y^1 and define $x^1 = x^*(y^1)$. This change will not affect our analysis.)

Definition 4.1

(i) Suppose that $\{x^n, y^n\}$ converges to (x^∞, y^∞). Then (x^∞, y^∞) will be called a pair of *stable strategies*. (Note that (x^∞, y^∞) is depending on the initial strategy x^1.)

(ii) Suppose that (x^∞, y^∞) is the limiting strategy pair, *independent* of the initial strategy x^1. Then (x^∞, y^∞) will be called a pair of *absolutely stable strategies*.

The following will simplify our analysis and help our understanding of stability.

Theorem 4.1 Suppose that $\{x^n, y^n\}$ converges to (x^∞, y^∞) and that $x^*(y)$ and $y^*(x)$ are continuous in a neighborhood of y^∞ and x^∞, respectively. Then $x^*(y^\infty) = x^\infty$ and $y^*(x^\infty) = y^\infty$. [Thus (x^∞, y^∞) is an intersection point of the graph of $x^*(y)$ and that of $y^*(x)$.] Furthermore, the following three statements are equivalent.

(A) $\{x^n, y^n\}$ converges to (x^∞, y^∞).

(B) $\{x^n\}$ converges to x^∞.

(C) $\{y^n\}$ converges to y^∞.

Proof By hypothesis $\{x^n\} \to x^\infty$ and $\{y^n\} \to y^\infty$. Since $y^{n+1} = y^*(x^n)$ and $x^{n+1} = x^*(y^n)$, the continuity of x^* and y^* implies $y^\infty = y^*(x^\infty)$ and $x^\infty = x^*(y^\infty)$. Finally, by continuity and the definition of $\{x^n\}$ and $\{y^n\}$ we see that whenever (B) [or (C)] holds then (C) [or (B)] holds; (B) and (C) together imply that (A) holds. On the other hand, when (A) holds, we know that (B) and (C) hold by the definition of convergence.

Q.E.D.

The following two examples help demonstrate the stability concept.

Example 4.1 In Figure 4 (x^0, y^0) is the unique pair of equilibrium strategies. Note that (x^0, y^0) is a pair of stable strategies if and only if the initial point $x^1 = x^0$. Hence it is not a pair of absolutely stable strategies. Note that if $x^1 \neq x^0$, then the optimal response strategies are divergent (the response strategies finally alternate between 0 and D_1 for Company 1 and between 0 and D_2 for Company 2). Thus, the existence of unique pair of equilibrium strategies does not imply the existence of stable strategies for arbitrary initial strategies. The first part of Theorem 4.1 says that a pair of stable strategies is also a pair of equilibrium strategies. However, the equilibrium pair (x^0, y^0) need be stable only for initial point $x^1 = x^0$ and hence need not be absolutely stable.

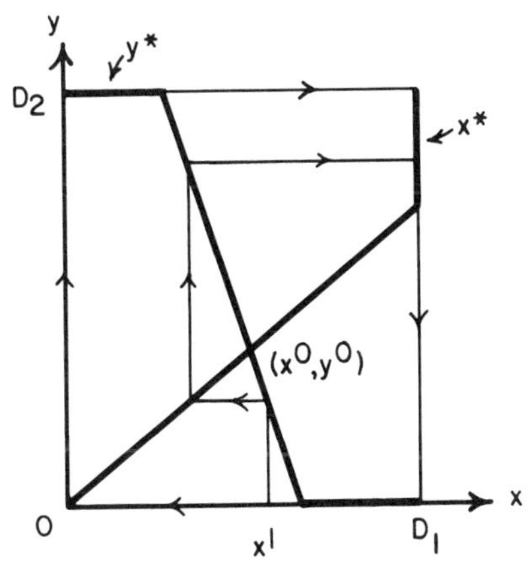

FIGURE 4

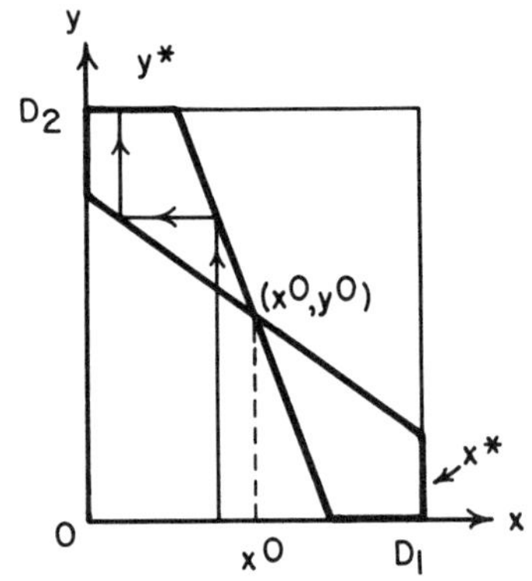

FIGURE 5

Example 4.2 In Model 2, if $e > 0$ and $c_1 = c_2 = 0$, one could obtain optimal response strategies (by selecting suitable parameters) as those depicted in Figure 5. Note that $(0, D_2)$, $(D_1, 0)$, and (x^0, y^0) are three pairs of equilibrium strategies. Suppose that the initial point $x^1 < x^0$. Then the sequence of response strategies $\{x^n, y^n\}$ converges to $(0, D_2)$. When $x^1 > x^0$, the sequence $\{x^n, y^n\}$ converges to $(D_1, 0)$. Thus $(0, D_2)$ and $(D_1, 0)$ are stable strategies for $x^1 < x^0$ and $x^1 > x^0$, respectively. But neither is a pair of absolutely stable strategies. This example shows that stable strategies may not be unique and that stable strategies may not be absolutely stable.

We now derive some results concerning stable strategies.

Theorem 4.2 Given an initial strategy x^1, suppose that the sequence of the response strategies $\{x^n, y^n\}$ is contained in $X \times Y \subset [0, D_1] \times [0, D_2]$, where X and Y are connected closed intervals. Suppose that $x^*(y)$ and $y^*(x)$ are continuous and have piecewise continuous derivatives and the following holds almost everywhere in $X \times Y$:

$$\left|\frac{dx^*(y)}{dy}\,\frac{dy^*(x)}{dx}\right| \leq r < 1$$

Then $\{x^n, y^n\}$ converges to $(x^\infty, y^\infty) \in X \times Y$.

Proof By hypothesis and the definition of $\{x^n\}$,

$$x^{n+1} - x^n = \int_{x^{n-1}}^{x^n} \frac{d}{dx} x^*(y^*(x))\, dx$$

$$= \int_{x^{n-1}}^{x^n} \frac{dx^*}{dy}\,\frac{dy^*(x)}{dx}\, dx$$

$$\leq r|x^n - x^{n-1}|$$

Thus

$$|x^{n+1} - x^n| \leq r|x^n - x^{n-1}| \tag{12}$$

Thus, by contraction mapping theory[2] (for instance, see [5]), $\{x^n\}$ converges to a point x^∞ in X. By invoking Theorem 4.1, the proof is completed. Q.E.D.

Theorem 4.3 Suppose that $x^*(y)$ and $y^*(x)$ are continuous and have piecewise continuous derivatives in $[0, D_2]$ and $[0, D_1]$, respectively, and

$$\left|\frac{dx^*(y)}{dy}\,\frac{dy^*(x)}{dx}\right| \leq r < 1$$

holds almost everywhere in $[0, D_1] \times [0, D_2]$. Then there exists a unique *absolute* stable strategy in $[0, D_1] \times [0, D_2]$.

---

[2]Another way to see the convergence is to note that since $r < 1$, (12) implies that $\{x^n\}$ is a Cauchy sequence; thus, it converges.

Proof The proof is similar to that of Theorem 4.2. However, in our current situation, $X = [0, D_1]$ and $Y = [0, D_2]$; hence contraction mapping theory assures the existence of a unique pair of stable strategies. Q.E.D.

Applying Theorem 4.3 and Theorems 3.2 and 3.3, we obtain

Theorem 4.4 In Model 2 with $e = 0$, suppose that the following hold:

(i) $c_1^2 < (b_1 - c_1)(b_2 + c_1)$

(ii) $c_1c_2 < (b_1 - c_1)(b_2 - c_2)$

(iii) $c_2^2 < (b_1 + c_2)(b_2 - c_2)$

or

(iv) $2c_1 < b_1$

(v) $2c_2 < b_2$

Then the pair of equilibrium strategies is the pair of absolutely stable strategies.

Proof In view of Theorem 4.3, we need to verify that

$$\left|\frac{dx^*(y)}{dy}\,\frac{dy^*(x)}{dx}\right| \leqq r < 1$$

almost everywhere. Since the derivative of a constant is zero, we need to verify only for those combinations: $(\bar{x}^*, \bar{y}^*)$, $(\bar{x}^*, \underline{y}^*)$, $(\underline{x}^*, \bar{y}^*)$, and $(\underline{x}^*, \underline{y}^*)$ which are defined in Theorems 3.2 and 3.3.

$$\frac{dx^*}{dy}\,\frac{dy^*}{dx} = \frac{-c_1}{b_1 - c_1}\,\frac{c_1}{b_2 + c_1} \quad \text{if } (x^*,y^*) = (\bar{x}^*,\bar{y}^*) \tag{13}$$

$$\frac{dx^*}{dy}\frac{dy^*}{dx} = \frac{-c_1}{b_1 - c_1}\frac{-c_2}{b_2 - c_2} \quad \text{if } (x^*,y^*) = (\bar{x}^*,\underline{y}^*) \tag{14}$$

$$= \frac{c_2}{b_1 + c_2}\frac{c_1}{b_2 + c_1} \quad \text{if } (x^*,y^*) = (\underline{x}^*,\bar{y}^*) \tag{15}$$

$$= \frac{c_2}{b_1 + c_2}\frac{-c_2}{b_2 - c_2} \quad \text{if } (x^*,y^*) = (\underline{x}^*,\underline{y}^*) \tag{16}$$

By setting $|\frac{dx^*}{dy}\frac{dy^*}{dx}| < 1$, (13) reduces to (i), (14) reduces to (ii), and (16) reduces to (iii), while (15) is always satisfied (since $c_2 < b_1 + c_2$ and $c_1 < b_2 + c_1$). This completes the first half of the proof. It remains to show that (iv) and (v) imply (i)-(iii). Observe that

$$2c_1 < b_1 \quad \text{iff} \quad c_1 < b_1 - c_1 \tag{17}$$

Since $0 < c_1 < b_2 + c_1$, (17) implies that (i) holds. Similarly,

$$2c_2 < b_2 \quad \text{iff} \quad c_2 < b_2 - c_2 \tag{18}$$

which implies that (iii) holds (because $c_2 < b_1 + c_2$). In addition, multiplying (17) with (18) on both sides, we obtain (ii). Q.E.D.

The result of Theorem 4.4 will be used in Example 5.1 of the next section.

§5. INCENTIVE FOR COOPERATION, CONSPIRACY, OR MERGER

In the previous two sections we used optimal response strategies to obtain equilibrium solutions which, under certain conditions, enjoy (absolute) stability. The main axiom behind the approach is the *individualism theme*--each company should try to maximize its own self-interested goal--sales. Each company is assumed to behave selfishly but not necessarily maliciously, as would be the case if each attempted to minimize the other company's objective. Real competitive behavior may actually follow this description to some extent because of anti-trust law constraints which prohibit most coordination or conspiracy between the two competing companies.

In this section we shall consider the question as to whether or not there is any incentive for cooperation, conspiracy, or merger between the two companies.

Suppose that (x^0, y^0) is a pair of equilibrium strategies, stable or not. Then the corresponding sales for the two companies are $\hat{s}_1(x^0, y^0)$ and $\hat{s}_2(x^0, y^0)$, respectively. Observe that x^0 and y^0 are obtained by maximizing $\hat{s}_1$ and $\hat{s}_2$, respectively, over $x \in [0, D_1]$ and $y \in [0, D_2]$.

Suppose that, through joint coordination, there is a pair of strategies $(\bar{x}, \bar{y})$ such that

$$\hat{s}_1(\bar{x},\bar{y}) \geqq \hat{s}_1(x^0,y^0) \tag{19}$$

and

$$\hat{s}_2(\bar{x},\bar{y}) \geqq \hat{s}_2(x^0,y^0) \tag{20}$$

hold simultaneously and with at least one holding strictly. (Thus, $[x^0, y^0]$ is not Pareto optimal.) Then there is an incentive for cooperation, conspiracy, or merger, since by redistributing their joint payoff $\hat{s}_1 + \hat{s}_2$ both companies may be better off. The larger the differences in (19) and (20), the stronger is the incentive. Of course, finding a "fair" redistribution law is very important in cooperation and constitutes a problem of its own. Such problems have been discussed in text books. (For instance, see [6,9,10].) For a recent summary, see Bergstresser and Yu [2]. We shall not elaborate on it here.

The following example demonstrates that in partially cooperative and partially competitive games, even if there exists the pair of equilibrium strategies which is absolutely stable, there still exists incentive for cooperation, conspiracy, or merger.

Example 5.1 In Model 2, let $a_1 = a_2 = 0.3$, $b_1 = 0.015$, $b_2 = 0.012$, $c_1 = 0.006$, $c_2 = 0.005$, $d = 0.2$, $e = 0$, $D_1 = 10$, $D_2 = 15$, $s_1 = 0.7$, $s_2 = 0.3$, and

$$[m_{ij}] = \begin{bmatrix} 0.8 & 0.4 \\ 0.5 & 0.6 \end{bmatrix}$$

Thus, $S = 1$, $q_1 = 0.71$, $q_2 = 0.46$,

$$f = 0.2\ (25 - x - y)$$

$$g = 0.3x - 0.015x^2 - (0.3y - 0.012y^2) + \begin{cases} 0.006(x - y)^2 & \text{if } x \geqq y \\ -0.005(x - y)^2 & \text{if } x < y \end{cases}$$

$$\hat{s}_1(x,y) = 0.71(1 + f(x,y)) + g(x,y)$$

$$\hat{s}_2(x,y) = 0.46(1 + f(x,y)) - g(x,y)$$

Note that the example is consistent with the following scenario.

Company 1 is the current market leader, enjoying a market share of 0.7 as opposed to Company 2's 0.3 share. If no advertising occurs, then in the next period Company 1 will remain the leader with a market share $q_1/(q_1 + q_2) = 0.61$.

Advertising budgets of \$10 (the units may be in thousands) for Company 1 and \$15 for Company 2 have been set, indicating that Company 2 is departing upon a somewhat aggressive advertising campaign. The firms' product advertising effectiveness is comparable (note $a_1 = a_2$) although Company 1 will suffer from saturation of the media with diminishing marginal returns somewhat earlier than Company 2 (note $b_1 > b_2$). If Company 2 allows Company 1 to outspend it on brand advertising (i.e., $y < x$), Company 2 will suffer more from the effect of Company 1's appearing to be the competitive leader than Company 1 would suffer if the situation were reversed (note that $c_1 > c_2$). This seems reasonable if we admit that Company 1 has probably become somewhat entrenched in the minds of consumers in the past, and this image is carrying forward.

Applying Theorems 3.2 and 3.3, we readily obtain (by direct substitution) the following optimal response strategies (note that $w_1 = 0.158$ and $w_2 = 0.208$, and case (ii) of Theorems 3.2 and 3.3 apply in this example). (See Figure 6.)

$$x^*(y) = \begin{cases} \min\{\bar{x}^*(y), 10\} = \bar{x}^*(y) & \text{for } y \in [0,\ 158/30] \\ \min\{\underline{x}^*(y), 10\} = \underline{x}^*(y) & \text{for } y \in [158/30,\ 15] \end{cases}$$

where

$$\bar{x}^*(y) = \frac{-2}{3}\,y + \frac{158}{18} \qquad \underline{x}^*(y) = \frac{1}{4}\,y + \frac{15.8}{4}$$

and

$$y^*(x) = \begin{cases} \min\{\underline{y}^*(x), 15\} & \text{for } x \in [0,\ 208/24] \\ \min\{\bar{y}^*(x), 15\} & \text{for } x \in [208/24,\ 10] \end{cases}$$

where

$$\underline{y}^*(x) = -\,\frac{5}{7}\,x + \frac{208}{14} \qquad \bar{y}^*(x) = \frac{1}{3}\,x + \frac{208}{42}$$

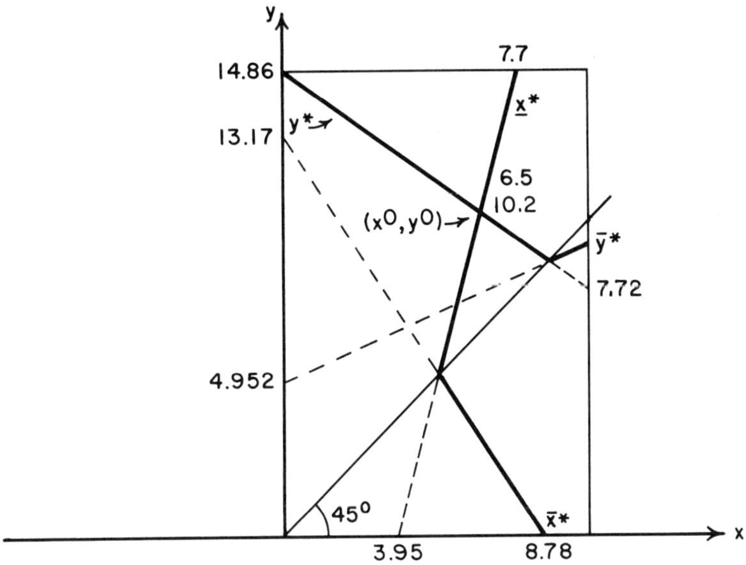

FIGURE 6

The unique pair of equilibrium strategies (see Figure 6) is given by

$$(x^0, y^0) = \left(\frac{429.2}{66}, \frac{858.4}{33}\right) \simeq (6.503, 10.212)$$

By invoking Theorem 4.4 (iv and v), we know that (x^0, y^0) is the pair of absolutely stable strategies. For (x^0, y^0) the payoff to both companies is $(\hat{s}_1, \hat{s}_2) = (1.322, 1.787)$. However, if $(x', y') = (5, 8)$ is used [thus $(5, 7) = (D_1 - 5, D_2 - 8)$ is applied to product strategies], then $(\hat{s}_1, \hat{s}_2) = (1.862, 2.116)$. Note the payoff of (x', y') is better than the payoff of (x^0, y^0) for both companies even if the former is not a pair of equilibrium or stable strategies. Note that this example can be interpreted as a continuous version of well known "Prisoner's Dilemma." Clearly, there may be incentive for both companies to cooperate, conspire, or merge.

§6. EXTENSION AND CONCLUSION

We have proposed two models to study partially cooperative and partially competitive marketing expansion games. Optimal response strategies, equilibrium strategies and stable strategies are derived and discussed. The incentive for cooperation, conspiracy, or merger is also briefly discussed. Clearly, our research is far from complete. Many gaming situations embody the features of partial cooperation and partial competition. For instance, in election campaigns, encouraging voters to vote is cooperative and getting as many votes for a particular candidate as possible is competitive. In industry, both management and labor unions want to increase sales, but each wants to maximize its share of the revenue. Whether or not our models can be used to study such conflicting situations is a topic for further research. Furthermore, we have only discussed games within the context of two companies. How do we extend the models for general n-company cases? And instead of the static cases used presently, can our models be extended to dynamic cases[3] as in differential games? These are just a few of the important questions which might be answered with further research.

---

[3]For such an extension, besides [14-15], the reader may be benefited by an excellent review on optimal control formulations in advertisement by Sethi [16].

APPENDIX A: A PROOF TO THEOREM 3.2

Recall that we replace x_1 and y_1 by x and y, respectively. In view of Theorem 3.1, we replace x_2 and y_2, respectively, by $D_1 - x$ and $D_2 - y$. Thus, equations (3), (5), and (6) reduce to

$$\hat{s}_1(x,y) = (1 + f(D_1 - x, D_2 - y))q_1 + Sg(x,y) \tag{A1}$$

where

$$f(D_1 - x, D_2 - y) = d(D_1 - x + D_2 - y) \tag{A2}$$

$$g(x,y) = (a_1x - b_1x^2) - (a_2y - b_2y^2) + \begin{cases} c_1(x - y)^2 & \text{if } x \geqq y \\ -c_2(x - y)^2 & \text{if } x < y \end{cases} \tag{A3}$$

By taking the derivative of $\hat{s}_1$ defined by (A1)-(A3) with respect to x and setting it equal to zero, the root of the resulting equation is given by

$$\bar{x}^*(y) = \frac{-c_1}{b_1 - c_1} y + \frac{w_1}{2S(b_1 - c_1)} \quad \text{if } x \geqq y \tag{A4}$$

or

$$\underline{x}^*(y) = \frac{c_2}{b_1 + c_2} y + \frac{w_1}{2S(b_1 + c_2)} \quad \text{if } x < y \tag{A5}$$

Since $x \in [0, D_1]$ we obtain the optimal response strategy $x^*(y)$ from (A4) and (A5):

If $x \geqq y$, then

$$x^*(y) = \begin{cases} 0 & \text{if } \bar{x}^*(y) < 0 \\ \bar{x}^*(y) & \text{if } 0 \leqq \bar{x}^*(y) \leqq D_1 \\ D_1 & \text{if } \bar{x}^*(y) > D_1 \end{cases} \tag{A6}$$

If $x < y$, then

$$x^*(y) = \begin{cases} 0 & \text{if } \underline{x}^*(y) < 0 \\ \underline{x}^*(y) & \text{if } 0 \leq \underline{x}^*(y) \leq D_1 \\ D_1 & \text{if } \underline{x}^*(y) > D_1 \end{cases} \tag{A7}$$

Observe that when $y = w_1/2Sb_1$, we have $\bar{x}^*(y) = \underline{x}^*(y) = y$. Thus the graph of $\bar{x}^*(y)$ and that of $\underline{x}^*(y)$ intersect at the line $x = y$ (see Figure 2), and thus $x^*(y)$ is continuous in y.

Three possible cases need to be considered.

Case 1 $w_1 \leq 0$. Thus $\bar{x}^*(y)$ and $\underline{x}^*(y)$ intersect at $y \leq 0$. Observe that $d\bar{x}^*/dy < 0$. Thus $\bar{x}^*$ can be used for optimal response strategy only when $y < 0$, which is impossible because $y \geq 0$, or when $y = 0$ at which point $\bar{x}^* = \underline{x}^*$. Thus, when $w_1 \leq 0$, $x^* > y$ never occurs and (A7) prevails. Observe that (A7) is identical to (7) and (8) of Theorem 3.2.

Case 2 $0 < w_1 < 2Sb_1D_1$. Thus $\bar{x}^*(y)$ and $\underline{x}^*(y)$ intersect at $w_1/2Sb_1 \in (0, D_1)$. Note that $0 < d\underline{x}^*/dy < 1$, and that $\bar{x}^*(y) \geq y$ whenever $y \leq w_1/2Sb_1$ and $\underline{x}^*(y) < y$ whenever $y > w_1/2Sb_1$. Thus, the optimal response strategy from (A6) and (A7) can be specified as in (9)-(11) of Theorem 3.2.

Case 3 $w_1 \geq 2Sb_1D_1$. Thus $\bar{x}^*(y)$ and $\underline{x}^*(y)$ intersect at $y = w_1/2Sb_1 \geq D_1$. Since $d\bar{x}^*/dy < 0$ and $d\underline{x}^*/dy > 0$, it can be verified that both $\bar{x}^* \geq D_1$ and $\underline{x}^* \geq D_1$ hold for $y \in [0, D_2]$. Thus $x^*(y) = D_1$ as specified in Theorem 3.2.

This completes the proof. Q.E.D.

APPENDIX B: OPTIMAL RESPONSE STRATEGIES FOR MODEL 2 WHEN $e > 0$

We start with $x^*(y)$, the optimal response strategy for Company 1. Using the notation of Appendix A, we have

$$\hat{s}_1(x,y) = (1 + f(D_1 - x, D_2 - y))q_1 + Sg(x,y)$$

where

$$f(D_1 - x, D_2 - y) = d(D_1 - x + D_2 - y) - e(D_1 - x + D_2 - y)^2$$

$$g(x,y) = (a_1 x - b_1 x^2) - (a_2 y - b_2 y^2) + \begin{cases} c_1(x - y)^2 & \text{if } x \geqq y \\ -c_2(x - y)^2 & \text{if } x < y \end{cases}$$

Let $D = D_1 + D_2$. By taking the derivative of $\hat{s}_1$ with respect to x and setting it equal to zero, the root of the resulting equation is given by:

$$\bar{x}^*(y) = \frac{-c_1 S - eq_1}{(b_1 - c_1)S + eq_1} y + \frac{w_1 + 2eq_1 D}{2(b_1 - c_1)S + 2eq_1} \quad \text{if } x \geqq y \qquad \text{(B1)}$$

$$\underline{x}^*(y) = \frac{c_2 S - eq_1}{(b_1 + c_2)S + eq_1} y + \frac{w_1 + 2eq_1 D}{2(b_1 + c_2)S + 2eq_1} \quad \text{if } x < y \qquad \text{(B2)}$$

where $w_1 = a_1 S - dq_1$. Since $x \in [0, D_1]$, we obtain the optimal response strategy $x^*(y)$ as follows:

If $x \geqq y$, then

$$x^*(y) = \begin{cases} 0 & \text{if } \bar{x}^*(y) < 0 \\ \bar{x}^*(y) & \text{if } 0 \leqq \bar{x}^*(y) \leqq D_1 \\ D_1 & \text{if } \bar{x}^*(y) > D_1 \end{cases}$$

If $x < y$, then

$$x^*(y) = \begin{cases} 0 & \text{if } \underline{x}(y) < 0 \\ \underline{x}^*(y) & \text{if } 0 \leqq \underline{x}^*(y) \leqq D_1 \\ D_1 & \text{if } \underline{x}(y) > D_1 \end{cases}$$

where $\bar{x}^*(y)$ and $\underline{x}^*(y)$ are given in (B1) and (B2). Observe that when

$$y = \frac{w_1 + 2eq_1D}{2b_1S + 4eq_1} \qquad \bar{x}^* = \underline{x}^* = y$$

That is, $\bar{x}^*(y)$ and $\underline{x}^*(y)$ again intersect at the line $x = y$. Thus, the optimal response strategies are well defined as above.

Although one can derive results similar to those of Theorem 3.2, we shall not stop to do so because of the complicated notation. The reader may observe that the optimal response strategy $x^*(y)$ is piecewise linear and similar to the case when $e = 0$. Clearly, when $e = 0$, the optimal response strategy reduces to that specified in Theorem 3.2.

Similarly, the optimal response strategy for Company 2 can be specified as follows:

If $x \geqq y$, then

$$y^*(x) = \begin{cases} 0 & \text{if } \bar{y}^*(x) < 0 \\ \bar{y}^*(x) & \text{if } 0 \leqq \bar{y}^*(x) \leqq D_2 \\ D_2 & \text{if } \bar{y}^*(x) > D_2 \end{cases}$$

where

$$\bar{y}^*(x) = \frac{c_1S - eq_2}{(b_2 + c_1)S + eq_2} y + \frac{w_2 + 2eq_2D}{2(b_2 + c_1)S + 2eq_2}$$

and $w_2 = a_2S - dq_2$.

If $x < y$, then

$$y^*(x) = \begin{cases} 0 & \text{if } \underline{y}^*(x) < 0 \\ \underline{y}^*(x) & \text{if } 0 \leqq \underline{y}^*(x) \leqq D_2 \\ D_2 & \text{if } \underline{y}^*(x) > D_2 \end{cases}$$

where

$$\underline{y}^*(x) = \frac{-c_2S - eq_2}{(b_2 - c_2)S + eq_2} y + \frac{w_2 + 2eq_2D}{2(b_2 - c_2)S + 2eq_2}$$

Again, $\bar{y}^*(x)$ and $\underline{y}^*(x)$ intersect at the line $x = y$ when

$$x = \frac{w_2 + 2eq_2D}{2b_2S + 4eq_2}$$

and the optimal response strategy $y^*(x)$ is well defined, and is piecewise linear.

ACKNOWLEDGMENT

This research was partially supported by a grant to the second author from the University General Research Fund of University of Kansas.

REFERENCES

1. Aaker, D. A., and Myers, J. G., Advertising Management, Prentice-Hall, Englewood Cliffs, New Jersey, 1975.

2. Bergstresser, K., and Yu, P. L., Domination Structures and Multicriteria Problems in N-Person Games, Theory and Decision, Vol. 8, 5-48 (1977).

3. Friedman, L., Game Theory Models in the Allocation of Advertising Expenditures, Operations Research, Vol. 6, 699-709 (1958).

4. Glicksburg, I. L., A Further Generalization of the Kakutani Fixed Point Theorem with Application to Nash Equilibrium Points, Proceedings of the American Mathematical Society, Vol. 3, 1952.

5. Korevaar, J., Mathematical Methods, Vol. 1, Academic Press, New York, 1968.

6. Luce, R. D., and Raiffa, H., Games and Decisions, John Wiley and Sons, New York, 1957.

7. Maffei, R. B., Brand Preferences and Simple Markov Processes, Operations Research, Vol. 8, 210-218 (1960).

8. Nikaido, H., and Isoda, K., Note on Non-cooperative Convex Games, Pacific Journal of Mathematics, Vol. 5, 807-815 (1955).

9. Owen, G., Game Theory, Saunders, Philadelphia, 1968.

10. Rapoport, A., N-Person Game Theory, University of Michigan Press, Ann Arbor, Michigan, 1969.

11. Shane, H. D., Mathematical Models for Economic and Political Advertising Campaigns, Operations Research, Vol. 25, 1-14 (1977).

12. Simon, J. L., Are There Economies of Scale in Advertising? Advertising Research, Vol. 5, 15-20 (1965).

13. Wrather, C., A Multi-type Branching Process Model for Predicting Future Automobile Population Sizes and Weight Distributions, Clemson University Review of Industrial Management and Textile Science, Spring 1977, pp. 65-86.

14. Clemhout, S., Leitmann, G., and Wan, H. Y., A Differential Game Model of Duopoly, Econometrica, Vol. 39, No. 6 (November 1971).

15. Leitmann, G., and Schmitendorf, Profit Maximization Through Advertising: A Nonzero Sum Differential Game Approach (working paper), 1978.

16. Sethi, S. P., Optimal Control Problems in Advertising: A Survey, SIAM Review, Vol. 19, No. 4 (October 1977).

STOCHASTIC CONTROL AND GAME ASPECTS IN A BIRTH-DEATH MODEL FOR SELLING ASSETS WITH A DEADLINE

Howard Blum

Graduate School of Business Administration
Rutgers University
Newark, New Jersey

§1. INTRODUCTION

We consider first a stochastic control model for selling a set of identical assets by a specified deadline and derive the optimal policy for the seller. Then we introduce game aspects and describe the impact of buyer behavior on the seller's policy.

The stochastic nature of the problem enters in the manner in which offers to buy the assets and cancellations of purchases arise. We study the viewpoint of the seller whose objective is to maximize the expected revenue at the deadline. Control in the model is exercised by the seller's acceptance or rejection of each offer to buy when it occurs. We will use a dynamic programming approach which leads to differential equations for an expected potential revenue function which is the expected revenue still to be earned, assuming an optimal policy is followed.

Offers to buy the assets are modeled as arising from each of several classes of customers at random points in time.

Each class of customers is characterized by the price offered, the average rate at which offers arise, the average rate at which cancellations occur, and the amount refunded when a cancellation occurs. The optimal policy in terms of the potential function is shown to be the following simple intuitive result: "Accept an offer if the price offered is at least equal to the decrease in the potential revenue function which would result."

Reservations systems are an example of selling assets with a deadline. Models similar to our stochastic control model have been used to study booking level policies for a single class of customers. Kosten [3] used a continuous model to study overbooking in airline reservations. Rothstein [6,7] introduces a constraint on the expected proportion of oversales at the deadline in a discrete version of Kosten's model in order to study overbooking in airline and hotel reservations. Liberman and Yechiali [8] derive the optimal policy in a discrete time, single class model where the seller can not only accept or reject randomly occurring offers but can also deterministically adjust the booking level at a cost. Our model can be used to study reservation and overbooking policies extended to more than one class of customers. In addition we consider the impact of buyer behavior in a game setting.

Our model specialized to selling a single asset to several classes of customers is a form of optimal stopping problem [4] such as the "secretary" problem [4,5]. The secretary problem consists of hiring the 'best' applicant during a sequence of interviews. The decision of whether to hire each applicant must be made prior to interviewing another applicant.

The usual criterion in stopping problems is to maximize the probability of choosing a largest element from the sequence available. Here we maximize the expected sum of the values of the set of chosen elements from a random process of known distribution and include the possibility of attrition by cancellations.

In the final section we introduce buyer behavior to create a game setting. After discussing possible decisions and information available to the buyers we turn to the impact of buyer behavior on the seller's policy. In the stochastic control model the seller's policy is shown to have the property described as final discount selling. We show that in a game setting, if switching of offers by buyers is not precluded, then the nature of the seller's policy should be one of initial discount selling rather than final discount selling.

§2. MODEL AND NOTATION FOR STOCHASTIC CONTROL

We will model the number of sales in effect to each class of customers as one component of a multidimensional birth and death process [1,2]. Hence we will be assuming:

1. Offers arise from each class in accordance with a poisson process, possibly nonstationary.
2. The acceptance or rejection of an offer does not effect the future rate of offers.
3. Cancellations from each class occur as a poisson process with average rate proportional to the number of sales in effect to the class.
4. Each sale or cancellation is for a single asset.

5. The amount refunded for a cancellation can be determined from the class of the individual being refunded and the time remaining to the deadline.

Assumptions 1 through 3 assert the Markov nature of the model which has discrete values in continuous time. This rules out certain intelligent behavior by customers, including new offers by rejected customers. The assumption of transitions only to neighboring values in 4 specializes the model to a birth and death process. The state of the model will be taken to consist of the number of sales in effect to each class, i.e., the components of the birth and death process. Assumption 5 assures that expected potential revenue, which is effected by future refunds, is conditionally independent of the past history of sales and refunds given the current state. This rules out, for example, refunds dependent upon the time separating purchase and cancellation. Note that if prices are time dependent but assumption 5 is satisfied, it is still possible to determine the potential revenue from knowledge of the state although it will not be possible to determine current revenue and hence the expected optimal revenue from the state.

The following notation is used:

c	number of assets to be sold
t	time remaining until the deadline
r	arrival rate of offers
i	class of an offer, $i = 1, \ldots, m$
p_i	probability that an offer is from class i
q_i	price offered by class i
w_i	cancellation rate per sale in effect to class i

b_i	refund paid for a cancellation by class i
x_i	number of sales in effect to class i
x	$(x_1x_2 \ldots x_m)$ the state
X	set of feasible states
δ_i	(00...1...0) unit m vector
u	u(x,t) = policy, a subset of {1, 2, ..., m} denoting the classes of offers to accept
V(x,t)	optimal expected potential revenue

The rates r and w_i, as well as the probabilities p_i, may depend upon the time remaining. In addition, the prices q_i and refunds b_i may depend upon the time remaining, subject to the limitation of assumption 5 and its subsequent discussion. The time argument for these parameters is omitted and the arguments for x, u, and V will also frequently be omitted when possible.

To illustrate the notation for change of state, assume that with time t remaining the state of the system is x(t) and during an interval of length Δt a sale to a customer of class i is the only event which occurs. Then at the end of the interval

$$x(t - \Delta t) = x(t) + \delta_i \tag{2.1}$$

Similarly, for a cancellation

$$x(t - \Delta t) = x(t) - \delta_i \tag{2.2}$$

A policy consists of a subset of classes, generally depending upon the state x and time remaining t, indicating those classes from which an offer is to be accepted. A policy

Subtracting V(x,t) and dividing both sides by Δt and passing to the limit as Δt goes to zero we obtain the system of differential equations for the optimal expected potential revenue functions V(x,t), alternately denoted by V_x.

$$\frac{d}{dt} V_x = \max\{(-r \sum_u p_i - \Sigma w_i x_i)V_x + r \sum_u p_i(q_i + V_{x+\delta_i}) + \Sigma w_i x_i(-b_i + V_{x-\delta_i})\} \qquad x \in X \tag{3.2}$$

This set of differential equations may be infinite depending on the set of feasible states X. Boundary conditions for V(x,t) would normally be given as payment received or penalties incurred at the deadline, t = 0. For example, if no payments or penalties are imposed, then

$$V(x,0) = 0 \qquad x \in X \tag{3.3}$$

Collecting terms of (3.2) dependent upon u and assuming the rate of offers r is positive, then it is necessary and sufficient that the optimal policy satisfy

$$\max_u \sum_u (V_{x+\delta_i} - V_x + q_i)p_i \tag{3.4}$$

The maximization of this sum can be achieved by choosing u to include all nonnegative terms. Terms for which p_i vanishes may be ignored. A further limitation is that the sale to class i be admissible, i.e., $x + \delta_i$ be a feasible state. Hence an optimal policy is characterized by

$$u^* = \{i : q_i \geq V_x - V_{x+\delta_i} \quad \text{and} \quad x + \delta_i \in X\} \tag{3.5}$$

This characterization of acceptable offers is quite intuitive, saying that an offer for which the price meets or exceeds the decrease in expected potential revenue should be accepted provided the result of the sale is an allowable state of the system.

Note that continuity of V(x,t), which is a consequence of it being a solution to the differential equations (3.2), implies that for a sufficiently small time remaining to the deadline the optimality of accepting offers is determined by the boundary condition V(x,0). For example, if (3.3) holds, then any feasible offer with positive price is acceptable in some interval near the deadline. This may be interpreted as implying that a policy of selling at a discount, i.e., at less than the highest price, is always optimal when the time is sufficiently close to the deadline and assets remain unsold. This property of the optimal policy will be referred to as final discount selling.

§4. A GAME SETTING

The stochastic control model studied so far assumes no intelligence on the part of the buyers. In this section we present some possibilities for expanding the role of the buyers to create a game setting. In the stochastic control version, we have been able to characterize the optimal policy for the seller and have pointed out the property of final discount selling. Our presentation of the game setting is preliminary and does not go so far as to determine optimal strategies for the seller or the buyers. We will, however, point out possible behaviors by the buyers which, if not ruled out, imply that the seller's strategy would be drastically altered in the game setting by the elimination of final discount selling to be replaced by initial discount selling. As the term suggests, by initial discount selling we mean a seller policy by which acceptance of offers of less than the highest price occurs during an initial time period, i.e., when the time remaining until the deadline is greatest.

In order to expand the role of the buyers, let the random points in time at which offers were assumed to be made be interpreted instead as points at which a potential buyer is created. In the stochastic control model this potential buyer is given no information or choice but must immediately make a single offer in accordance with class i selected at random with probability p_i. The extension to a game model could afford the customer various choices including the choice of class of offer, the timing of an offer, and the possibility of more than a single offer including subsequent switching of classes. Costs to the potential buyer for making offers could be imposed. Note that selection of class of offer by the customer is the mechanism provided in our model for selecting the price of an offer and the conditions on cancellations.

The choice by the customer could be made in accordance with a utility function dependent upon factors including the probability of acquiring the asset, the price paid, and the restrictions on cancellations. The information available to the customer could include the time remaining until the deadline and the number of assets remaining to be sold. The composition by classes of the existing sales in effect would also be valuable information to the customer because of its influence on the distribution of future cancellations. The inclusion of stochastic factors in the utility function of buyers would result in nondeterministic behavior for a given information set, reflecting a diversity of potential buyers.

We now consider the behavior of potential buyers faced with a seller policy that includes final discount selling. One buyer strategy would be to repeatedly make offers at a discount price if he is aware that unsold assets remain as the deadline approaches. This would be akin to a standing order for a discount purchase. The information of a standing discount offer could be used to advantage by the seller in determining a policy. This phenomenon would not eliminate final discount selling but would reduce the period during which it would be performed to the point at the deadline unless there is attrition of offers to buy that are kept waiting.

A more troublesome possibility from the viewpoint of the seller is the intentional switching of a buyer from a full price to a discount purchase. A cautious buyer could first assure himself of an asset, if available, by first paying the full price, then repeatedly making offers at a discount price, and subsequently cancelling the original purchase if a discount offer is accepted. A variation of this involves cancellation of the full price offer and an immediate offer to buy at discount as the deadline approaches.

If such intentional switching is not precluded by rules of the game, then it is simple to show that the strategy of the seller should not include final discount selling. If the demand for assets at full price is not sufficient, then an optimal policy for the seller must contain some interval of discount selling. If this interval is preceded by an interval of selling only at full price, then switching will occur and negate its effect. Hence we conclude that if switching is not precluded then the policy of the seller should only contain an initial discount interval.

ACKNOWLEDGMENT

The author is happy to acknowledge helpful comments from Professor Fred Sudit of Rutgers, Mr. Herb Ayres of Morgan Guaranty, and Professor Gary Gottlieb of New York University.

REFERENCES

1. Feller, W., An Introduction to Probability Theory and Its Applications, Vol. 1, Wiley, 1950.
2. Cooper, R., Introduction to Queueing Theory, Macmillan, 1972.
3. Kosten, L., Een Mathematisch Model Voor Eeen Reserverings-probleem, Statistica Neerlandica, 14, 85-94 (1960).
4. Chow, Y., Robbins, H., and Siegmund, D., Great Expectations, The Theory of Optimal Stopping, Houghton Mifflin, 1971.
5. Gilbert, J., and Mosteller, F., Recognizing the Maximum of a Sequence, Amer. Stat. Assoc., Vol. 61, 35-73 (1966).
6. Rothstein, M., An Airline Overbooking Model, Transportation Science, Vol. 5, No. 2, 180-192 (May 1971).
7. Rothstein, M., Hotel Overbooking as a Markovian Sequential Process, Decision Sciences, Vol. 5, 389-404 (1974).
8. Liberman, V. L., and Yechiali, U., On the Hotel Overbooking Problem--An Inventory System with Stochastic Cancellations, Management Science, Vol. 24, No. 11 (July 1978).

A LIFE-CYCLE THEORY OF MARRIAGE AND DIVORCE:

A PARETO OPTIMAL DIFFERENTIAL GAME MODEL

Simone Clemhout

Cornell University
Ithaca, New York

ABSTRACT

Recently, economists have joined sociologists and psychologists in analyzing the importance of the family. From an economic point of view family decisions have pervasive influence on the economy thereby raising the question of policy intervention over the life cycle. This is particularly important as divorce rates are 40 percent and in an uptrend.

However, it remains to be shown whether or not pro-marriage legislations are socially desirable. This paper demonstrates that there are cases where in the absence of children, marriage followed by divorce is Pareto Optimal. In such cases legislation designed to perpetuate marriage is nonoptimal but legislations to compensate inequities in settlements are eminently desirable from a social point of view.

A differential game model is analyzed and comparative dynamics or sensitivity analysis characterizes people's behavior and determines whether or not they will marry and if they marry will they stay married for life or divorce.

§1. INTRODUCTION

Long ago psychologists and sociologists focused on the family as a major area of endeavor. However, only recently with the onset of the "new home economics," as aptly coined by Marc Nerlove, have economists joined the field [1]. The pioneering and innovative work of the Chicago School, with Gary Becker as a vanguard, has established this area as a prominent one. Their work as well as that of others has demonstrated that marriage and marriage instability have crucial implications for fertility and population growth, the labor force participation of women, patterns of consumer expenditures (especially housing and consumer durables), inequality of income, ability and other characteristics among families [2]. Despite the valuable insights obtained, these views are static, while the problem of marital instability over the life cycle is essentially dynamic.

Time of marriage, interpersonal relationships and the power play and harmony within the family are instrumental in the process of family decision making [3,4]. But, of particular interest, is the increasing instability of marriage leading to separation and divorce. Divorce rates are now about 40 percent and seem to be in an uptrend, in the USSR as well as in the USA.

In view of the importance of the consequences of family decisions for the economy and questions of fairness and equity regarding the life-cycle flows of income for the parties before and after divorce, the emphasis has shifted from private families to concomitant public responsibilities. Thus, the interface of public decisions and impact on private families has become a major area of public policy [5].

The United States' Senate hearings on "American Families: Trends and Pressure," under the sponsorship of now Vice-President Mondale, signal the sponsorship of family-impact statements on federal legislative proposals. However, the idea of the family policy is wrought with ambiguity resulting from the absence of explicit theory. What are the functions of the family? Is it a biological unit for reproduction and rearing of children, is it an organizational convenience, does it serve an "asylum function" [5]? Family stability becomes an issue by breaking the life cycle into segments, each susceptible to government intervention.

What policy will be encouraged depends on what segments can be influenced and how the political system responds to what particular issues. Do we want to encourage stability or mitigate the consequences of marital instability? Congresswoman Yvonne Burke proposed the displaced homemaker assistance act, now an amendment to the Comprehensive Employment and Training Act of 1973 (CETA) legislation [6].

Although certain financial benefits are provided for certain women, the "target group dilemma" faces the displaced homemaker who cannot turn to the usual sources for temporary financial relief. Even though she is unemployed, she does not not qualify for unemployment compensation. Under present law, if she is divorced after having been married less than 20 years, she may not qualify for social security benefits. This may soon be changed to ten years under new legislation. If she has no children under age 18 she will be ineligible for aid to families with dependent children [7].

Marital instability and its consequences for certain target groups raise the question: "Is marriage worthwhile?" Is it the best situation for both partners? In other words, can marriage ever reflect a Pareto-Optimal situation. Whatever the answer, does the consideration of divorce change the picture?

The model proposed in this paper demonstrates that it is possible for marriage to be Pareto Optimal with respect to the interests of the two spouses. This is true for two distinct types of marriages. One type of marriage can be "until death do we part;" that is, the marriage ends with the death of one of the spouses. The other type is "marriage with contemplated eventual divorce;" that is, spouses when they get married do contemplate the possibility of divorce and yet enter into wedlock. Such a marriage may be "group rational" for them and divorce can be consistent with Pareto Optimality; it is not necessarily a result of unforeseen events.

A scenario of "marriage with contemplated eventual divorce" is as follows. Compared with remaining single, the spouses can maintain a higher rate of capital accumulation (be it real capital or human capital)--or a lower rate of capital decumulation--in marriage. Therefore, they enter into wedlock. Beyond a certain instant, attainable personal payoff (say happiness) may be higher by being single than remaining married, given the same stock of capital. So, at some instant in the course of the marriage phase of the life-cycle the spouses part company, returning to "singlehood" with a higher level of capital than would have been attainable without marriage. The cumulation (and decumulation) of wealth relates to the work-leisure choice of the couples. They may or may not leave a bequeath at death.

This model while based on simple assumptions captures certain essences of reality. The family functions as an economic, as well as a social unit. Its formation or dissolution affects the earning capacities of the spouses; thus, their wealth accumulation is also influenced. Wealth enters into the payoff functions of both spouses. It affords the consumption which is one of the determinants of "happiness;" the latter, of course, also depends upon the marital status of the individuals, parametrically. Without excluding other dimensions (e.g., noneconomic) of marriage, the present model sheds light on the economic motives of the individuals, and these among other factors are of prime importance in the life cycle.

The discontinuities in the payoff function over the life cycle relates to the literature on optimal control problems with integrand discontinuous with respect to time [8].

§2. THE DIFFERENTIAL GAME FRAMEWORK

Since the problem of marital instability over the life-cycle is dynamic and involves interactions between husband and wife, differential games is the natural approach of analysis. Since we inquire whether or not it is Pareto Optimal (or "group rational") to be married, but "not for life," the Pareto equilibrium concept is relevant in such a game.

It is known that if one can achieve the maximum of the weighted sum of individual utilities along a trajectory, that trajectory must be Pareto Optimum (i.e., a "non-inferior" solution) [9]. Hence, one asks whether there are cases where a trajectory maximizing the weighted sum of the utilities entices couples to stay married only for part of their life time.

During marriage the spouses accumulate wealth. Let

$$x = \begin{pmatrix} x_1 \\ x_2 \end{pmatrix} \tag{1}$$

be a pair of wealth positions for spouses 1 and 2. During marriage the couple's common property is $x_1 + x_2$. When divorce occurs x_1 and x_2 are their respective individual shares. u_i is the control for person i = 1, 2, which, for instance, could be the number of hours they work either in the market place or in the household.

The controlled system, for i = 1, 2, is described by the following set of differential equations:

$$\dot{x}_i = \begin{cases} f_i(x_1,x_2,t,u_1,u_2) & \text{(the wealth cumulative function in marriage)} \\ f^i(x_i,t,u_i) & \text{(the wealth cumulative function out of marriage)} \end{cases} \tag{2}$$

The payoff functions of the partners (i.e., happiness) at any point in time (for i = 1, 2) are:

$$\begin{cases} g_i(x_1,x_2,t,u_1,u_2) & \text{in marriage} \\ g^i(x_i,t,u_i) & \text{out of marriage (after divorce or death of spouse)} \end{cases} \tag{3}$$

t is time; t_i is the remaining life of spouse i at the date of marriage; $t \in [t_0,t_4]$, t_0 is the point when marriage starts; $t_4 = \min[t_1,t_2]$ is the time of death of one of the partners; and $t_3 < t_4$ is the time of divorce.

The scalarized Pareto differential game model becomes the control problem

$$\max_{\{u_i(\cdot)\}} w_1 \int_{t_0}^{t_3} g_1(x_1,x_2,t,u_1,u_2)dt$$

$$+ w_2 \int_{t_0}^{t_3} g_2(x_1,x_2,t,u_1,u_2)dt$$

$$+ w_1 \int_{t_3}^{t_1} g^1(x_1,t,u_1)dt + w_2 \int_{t_3}^{t_2} g^2(x_2,t,u_2)dt \qquad (4)$$

subject to

$$\begin{aligned} \dot{x}_i &= f_i(x_1,x_2,t,u_1,u_2) \qquad i = 1, 2 \qquad t_0 \le t \le t_3 \\ \dot{x}_i &= f^i(x_i,t,u_i) \qquad i = 1, 2 \qquad t > t_3 \end{aligned} \qquad (5)$$

$x_i(0)$ given, $x_i(t_i) = \bar{x}_i$, $i = 1, 2$. The condition of a fixed bequeath, $\bar{x}_i$, is postulated for simplicity. x_i and u_i may be vectors of appropriate dimensions.

One can assume that behavioral motivators are (i) that wealth accumulation is easier in marriage and (ii) that at the time of divorce enjoyment is higher being single than married.

§3. AN EXAMPLE

To show that marriage may end in divorce under Pareto Optimality, one has to characterize some Pareto Optimal trajectory implying divorce. This is done by solving a scalarized optimal control problem, viz., a trajectory maximizing the weighted sum of utility integrals is a Pareto Optimal trajectory [10,11].

In this example, for simplicity, we assume symmetric marriage, which implies that $w_1 = w_2 = 1$, $f_1(\cdot) = f_2(\cdot)$,

$f^1(\cdot) = f^2(\cdot)$, $g_1(\cdot) = g_2(\cdot)$, $g^1(\cdot) = g^2(\cdot)$, $t_1 = t_2 = T = 1$, and $x_1(0) = x_2(0)$. Supposing that the individual utility function is strictly concave, the consumption time path must be the same for both spouses.

The state equations are

$$\begin{aligned} \dot{x}_1 + \dot{x}_2 &= a^{(0)}(x_1 + x_2) - (c_1 + c_2) \quad \text{(in marriage)} \\ \dot{x}_i &= a^{(1)}x_i - c_i \quad \text{(in divorce)} \end{aligned} \tag{6}$$

where x_i and c_i are wealth and consumption for i = 1, 2, respectively.

We assume that $a^{(0)} > a^{(1)} > 0$, $c \triangleq a^{(1)}/a^{(0)} \in (0,1)$, and $a^{(j)}$ is the income-wealth ratio for phase j (with j = 0 in marriage, j = 1 after divorce).

Following the tradition established by Chakravarty [12] (see also [13]), the payoff functions are:

$$g_i = p^{(0)}c_i^{\ b} \qquad \text{and} \qquad g^i = p^{(1)}c_i^{\ b} \tag{7}$$

$b \in (0,1)$, b being the consumption elasticity of utility, $D \triangleq p^{(0)}/p^{(1)}$.

Consequently, the problem is equivalent to

$$\underset{\substack{\{c_i\} \\ 0 \le t_3 \le 1}}{\text{Max}} \quad J_i = \int_{t_0}^{t_3} p^{(0)}c_i^{\ b}\,dt + \int_{t_0}^{t_3} p^{(1)}c_i^{\ b}\,dt \tag{8}$$

subject to:

$$\begin{aligned} \dot{x}_i &= a^{(0)}x_i - c_i \qquad t_0 \le t \le t_3 \\ \dot{x}_i &= a^{(1)}x_i - c_i \qquad t_3 \le t \le 1 \end{aligned} \tag{9}$$

$x_i(0)$ given, $x_i(1)$ given. For simplicity, $x_i(1)$ is set to be 0 (no bequeath).

The subscript i shall be dropped below when there is no confusion.

The Hamiltonian format implies (1) for the period from the start of the marriage up to the time of divorce:

$$H^{(0)} = p^{(0)}c^b + y^{(0)}[a^{(0)}x - c] \qquad t_0 \le t \le t_3 \tag{10}$$

with conditions for maximization:

$$\dot{x} = a^{(0)}x - c$$

$$\dot{y} = -a^{(0)}y^{(0)}$$

$$0 = \frac{\partial H^{(0)}}{\partial c} = bp^{(0)}c^{b-1} - y^{(0)}$$

(2) for the post-divorce period:

$$H^{(1)} = p^{(1)}c^b + y^{(1)}[a^{(1)}x - c] \qquad t_3 \le t \le 1 \tag{11}$$

with conditions for maximation:

$$\dot{x} = a^{(1)}x - c$$

$$\dot{y} = -a^{(1)}y^{(1)}$$

$$0 = \frac{\partial H^{(1)}}{\partial c} = bp^{(1)}c^{b-1} - y^{(1)}$$

where the co-state variable y represents the shadow value of wealth in terms of utility (i.e., happiness).

The above two sets of conditions associated with the Hamiltonians $H^{(0)}$ and $H^{(1)}$ are necessary to maximize the "in-marriage" and "after-divorce" payoffs over these two respective phases. The maximization of the life-cycle payoff, combining the two phase specific payoffs, is solved in two stages:

(1) Provisionally fixing an arbitrary date t_3 for divorce, the sum of the two phase specific payoffs is maximized.

Since the initial wealth of the "after-divorce" phase equals the terminal wealth of the "in-marriage" phase, we solve the sequence of phase-specific optimization problems backward in time, according to Bellman's Principle of Optimality [8]. The link between the two phases is established through the transversality condition, namely:

$$y^{(0)}(t_3) = y^{(1)}(t_3) \tag{12}$$

The economic interpretation signifies that the terminal shadow value for wealth in the "in-marriage" phase, $y^{(0)}(t_3)$, is evaluated at the initial shadow value for wealth in the "after-divorce" phase, $y^{(1)}(t_3)$. Consequently, for each arbitrarily specified t_3, there corresponds an optimal value of $x_i(t_3)$, i.e., the level of wealth at divorce.

The maximum attainable life-cycle payoff (for spouse i), given the provisional time of divorce t_3, is

$$\begin{aligned}\hat{J}_i(t_3) &= x_i(0)^b e^{a_i^{(0)} b t_3}\left(\frac{1-b}{ba_i^{(1)}}\right)^{1-b}\left[\left(e^{a_i^{(1)} b(1-t_3)/(1-b)} - 1\right)\right. \\ &\quad \left. + D^{\frac{1}{1-b}} \frac{a_i^{(1)}}{a_i^{(0)}}\left(1 - e^{-a_i^{(0)} bt_3/(1-b)}\right)\right]^{1-b} \\ &= x_i(0)^b \left(\frac{1-b}{ba_i^{(1)}}\right)^{1-b}\left[e^{a_i^{(1)} b/1-b}\, e^{(a_i^{(0)} - a_i^{(1)})bt_3/(1-b)}\right. \\ &\quad \left. - \left(1 - \frac{a_i^{(1)}}{a_i^{(0)}} D^{\frac{1}{1-b}}\right) e^{a_i^{(0)} bt_3/(1-b)} - \frac{a_i^{(1)}}{a_i^{(0)}} D^{1(1-b)}\right]\end{aligned} \tag{12}$$

(2) $\hat{J}_i(t_3)$ depends on t_3 alone, thus it remains to maximize $\hat{J}_i(t_3)$ by selecting the optimal value of t_3. Since

t_3 is restricted to the unit interval [0,1], the Kuhn-Tucker complementary slackness condition [14] implies that three possibilities might occur:

Case 1 ("until death do we part"):

$$t_3^* = 1 \text{ when } \partial J/\partial t_3 > 0 \text{ at } t_3 = 1 \qquad (13)$$

Case 2 (marriage does not occur):

$$t_3^* = 0 \text{ when } \partial J/\partial t_3 < 0 \text{ at } t_3 = 0 \qquad (14)$$

Case 3 (marriage with subsequent divorce at t_3):

$$t_3^* \in [0,1] \text{ when } \partial J/\partial t_3 = 0 \qquad (15$$

A careful analysis of equation (12) and its derivatives will determine whether or not marriage occurs, if and when divorce results, and the duration of marriage, in terms of $p_i^{(0)}$ and $p_i^{(1)}$, which capture noneconomic parameters, as well as $a_i^{(0)}$ and $a_i^{(1)}$, the capacity to accumulate wealth.

It can be shown that if

$$p_i^{(1)} \leqq p_i^{(0)} \qquad \text{or} \qquad D = p_i^{(1)}/p_i^{(0)} \geq 1$$

i.e., marriage is superior to singlehood at all consumption levels, people always stay married for life, thanks to the superior accumulative capacity of the marriage state (i.e., $a_i^{(0)} > a_i^{(1)}$). Analytically, two situations may be distinguished.

(i) If

$$\left(\frac{p_i^{(0)}}{p_i^{(1)}}\right)^{\frac{1}{1-b}} \frac{a_i^{(1)}}{a_i^{(0)}} \geq 1 \qquad (16)$$

the $\hat{J}(t_3)$ is convex and strictly increasing in t_3 and the boundary solution $t_3 = 1$ (i.e., married for life) prevails.

(ii) Alternatively, if

$$\left(\frac{p_i^{(0)}}{p_i^{(1)}}\right)^{\frac{1}{1-b}} \frac{a_i^{(1)}}{a_i^{(0)}} < 1 \tag{17}$$

$\hat{J}(t_3)$ is strictly concave but still increasing in t_3, at most $\partial\hat{J}/\partial t_3 = 0$ at $t_3 = 1$, for the marginal case of $p_i^{(1)} = p_i^{(0)}$, again $t_3 = 1$ is optimal, and in both cases marriage is for life.

However, if

$$p_i^{(1)} > p_i^{(0)}$$

so that married life is less happier than singlehood, either the two persons will not get married, i.e., $t_3 = 0$, or if they do they will divorce each other at some future time. This can be proved from the fact that

$$\left.\frac{\partial\hat{J}}{\partial t_3}\right|_{t_3=1} = K_1(D^{1/1-b} - 1) < 0 \tag{18}$$

where K_1 is a complicated expression with a positive sign and $D = p_i^{(0)}/p_i^{(1)} < 1$.

Consequently, the necessary condition for the marriage-for-life case (case 1 above, Eq. (13)) to be optimal, is violated.

However, it is entirely possible that with equal levels of happiness not related to their economic situations, couples decide on their preference for being single or married on the basis of consumption levels in these two states, e.g.,

$$g_i(c_i) = (c_i)^{b(0)} \qquad g^i(c^i) = (c^i)^{b(1)} \qquad b(0) \neq b(1) \tag{19}$$

where b, the consumption elasticity of utility, is different for the two states of marriage and divorce. In this case there will be a cross-over between the two functions representing happiness in terms of consumption. It will be located at $c_i = 1 = c^i$. Thus, there will be a switch in the two states before and after the cross-over and in this manner marriage and divorce can also occur. This is true even if equation (18) holds.

To demonstrate and analyze the possibility of a Pareto-Optimal marriage-then-divorce case (case 3 above, equation (15)), we assume that $D = p_i^{(0)} < p_i^{(1)} < 1$, although equation (19) might also apply. To simulate reality more closely, more complicated models might be entertained, but for implicit policy considerations a simpler model is advantageous.

Whether a couple will get married to each other and whether when married they will stay married for some time depends on the equation

$$\left.\frac{\partial \hat{J}}{\partial t_3}\right|_{t_3 \in [0,1]} = K_2 \left[e^{a^{(1)} b(1-t_3)/(1-b)} - \frac{1 - \left(\dfrac{a_i^{(1)}}{a_i^{(0)}}\right) D^{\frac{1}{1-b}}}{1 - \left(\dfrac{a_i^{(1)}}{a_i^{(0)}}\right)} \right] = 0 \tag{20}$$

where K_2 is a complicated expression with positive sign.

The solution of equation (20) falls strictly between 0 and 1, if:

$$\frac{1-b}{b}\log\left[\frac{1-\dfrac{a_i^{(1)}}{a_i^{(0)}}D^{\frac{1}{1-b}}}{1-\dfrac{a_i^{(1)}}{a_i^{(0)}}}\right] < a^{(1)} \tag{21}$$

The violation of the above inequality means that no marriage would take place.

§4. COMPARATIVE DYNAMICS OR SENSITIVITY ANALYSIS

In this model parametric characterization decides whether people marry or not. How people find their partners, in other words, what search strategy they follow in pairing, is of no concern to us here. The subject has been discussed by Gary Becker and commented upon in Ref. 3.

Retaining the hypothesis of symmetry, each potential couple's behavior can be characterized by a vector of four parameters $[a, b, (a_i^{(1)}/a_i^{(0)}), D]$. For each vector one can determine the following possible occurrences:

1. Will they marry at all? The answer is yes in cases 1 and 3, and no in case 2 of Section 3.
2. Will a married couple divorce? The answer is yes in case 3 and no in case 1 of Section 3.
3. How soon will they divorce? Case 3 in Section 3 will determine the optimal time of divorce.

Comparative dynamic analysis describes how the date of divorce varies with the four parameters. Couples who will not divorce will have their parameter vectors located in some subset of the parametric space.

In principle, any change in the distribution of potential couples over the parametric space can be analyzed so as to reveal the socioeconomic impact in quantitative terms. This can be achieved with numerical simulation.

To decide which potential couple will ever marry and to deduce how the "propensity to marry" varies with the characteristics of a potential couple, we construct the parametric map in Figure 1. It reflects the information contained in the inequality in equation (21).

In Figure 1, for each given value of the consumption elasticity of happiness, b, plot on the vertical axis $a_i^{(1)}$, denoting the capacity to accumulate wealth when single, and on the horizontal axis the ratio $(a_i^{(1)}/a_i^{(0)})$ indicating the relative advantage of singlehood for accumulation. This ratio is assumed to be less than 1, supposing that marriage is favorable to wealth accumulation. The small diagram to the right of Figure 1 shows how to interpret the changes in the accumulative capacities plotted in the diagram. All potential couples sharing the same specified b are represented by particular points in the AOBC shaded region; each point indicates a pair of values $(a_i^{(1)}/a_i^{(0)}, a_i^{(1)})$. For those subclasses of couples exhibiting particular values not only for b but also for D (the degree of relative happiness in marriage), D will have a designated boundary line, OD, in the parametric space.

Couples whose pair of parameters $(a_i^{(1)}/a_i^{(0)}, a_i^{(1)})$ is located on or below their corresponding OD curve will not marry. Those with their parameter pair above OD will marry and then divorce, if $D < 1$. Given b and D, if a couple has its pair of parameters located exactly on OD, they will not

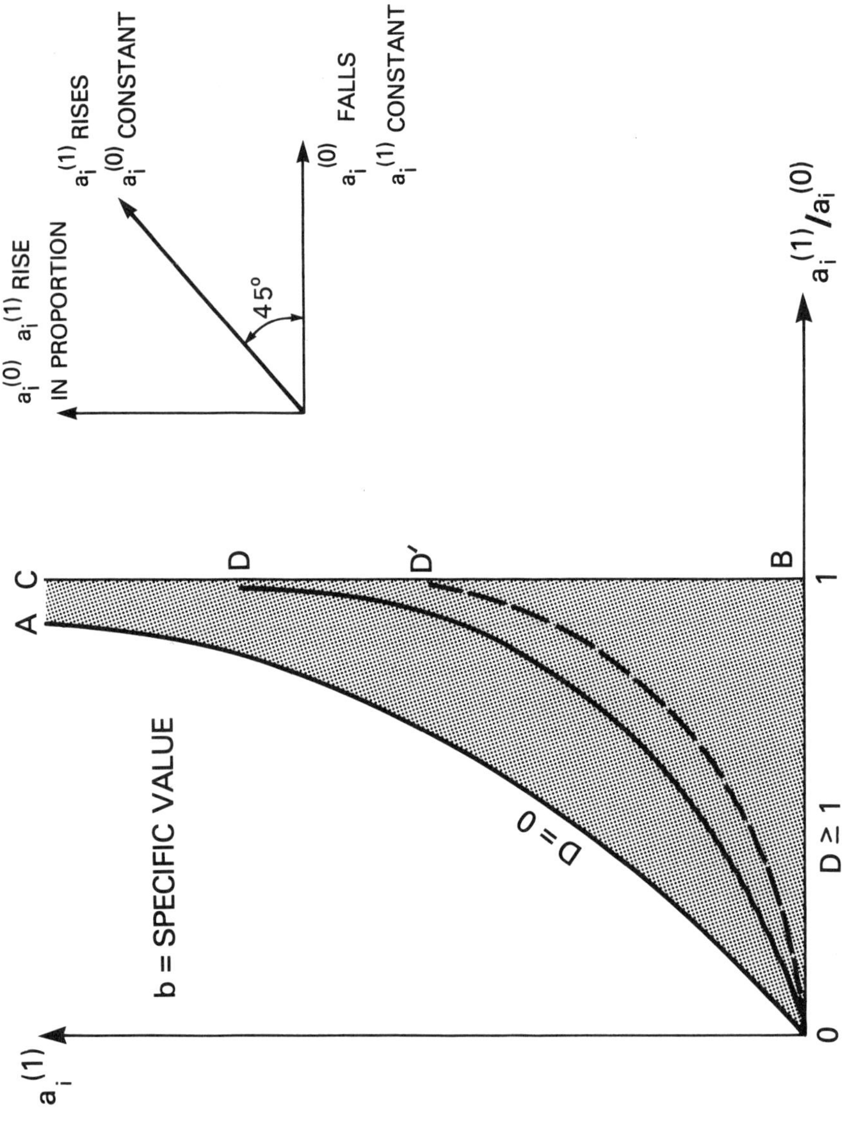
$a_i^{(1)}$
b = SPECIFIC VALUE
A
C
D
D′
B
D = 0
D ≥ 1
0
1
$a_i^{(1)}/a_i^{(0)}$
$a_i^{(0)}$ $a_i^{(1)}$ RISE IN PROPORTION
45°
$a_i^{(1)}$ RISES
$a_i^{(0)}$ CONSTANT
$a_i^{(0)}$ FALLS
$a_i^{(1)}$ CONSTANT

Figure 1

marry, while a couple with the same b and D, but greater cumulative capacities both during marriage and after divorce, will marry for some time.

Alternatively, if a third couple has the same b and D, as well as the same cumulative capacity in divorce as the first one, but greater cumulative capacity in marriage, they will marry for some time. A fourth couple with the same b, $a_i^{(1)}$, and $(a_i^{(1)}/a_i^{(0)})$ as the first couple, but enjoying married life relatively more (higher D, say D'), will have their corresponding boundary curve at OD' in Figure 1; consequently, they will marry for a period of time.

In Figure 1, if the value of D falls toward zero, the boundary curve approaches OA while when the value of D rises to 1 or beyond, it moves toward OB and stays there for all $D \geq 1$. The limit $(D = 0)$ means a happy-less marriage and no couple enters into wedlock. The limit $(D = 1)$ implies equal happiness, so that with cumulative capacity favoring married life, all will marry.

Focusing now on case 3 in Section 3, which represents the situation where marriage takes place and is followed by divorce, the optimal duration of marriage is

$$t_3^* = 1 - \frac{1 - b}{a_i^{(1)} b}\left[\log\left(1 - \frac{a_i^{(1)}}{a_i^{(0)}} D^{\frac{1}{1-b}}\right) - \log\left(1 - \frac{a_i^{(1)}}{a_i^{(0)}}\right)\right] \tag{22}$$

How does t_3^* vary with respect to parametric variations in $a_i^{(1)}$, $(a_i^{(1)}/a_i^{(0)})$, D, and b? Straightforward differentiation yields

$$\frac{\partial t_3^*}{\partial a_i^{(1)}} > 0 \tag{23}$$

$$\frac{\partial t_3^*}{\partial (a_i^{(1)}/a_i^{(0)})} < 0 \tag{24}$$

$$\frac{\partial t_3^*}{\partial D} > 0 \tag{25}$$

$$\frac{\partial t_3^*}{\partial b} \quad \text{(has ambiguous sign)} \tag{26}$$

In summary, the meaning of equations (23) to (26) is that for those who marry and then divorce, the duration of marriage lengthens if the capacities to cumulate in both the marriage and after divorce phases increase proportionately, if the capacity to cumulate in marriage alone increases, and if the relative happiness in marriage versus after divorce increases. The effect of a higher consumption elasticity to happiness defies easy characterization.

§5. POLICY RECOMMENDATIONS

The discussion in this paper demonstrates that within our assumptions promarriage legislations have little justification in themselves. It is clearly seen that marriages *where no children are present* can be followed by divorce and yet represent a Pareto Optimal situation. In other words, it would not be possible to improve the conditions of one person without damaging those of the other. These cases do not represent all marital situation, as we have seen, but may hold for a subset of cases.

If the after divorce settlements do not represent an equal flow of income (in our symmetric case), equity is violated and equity considerations require the policy maker to set guiding norms to settlements or to compensate the parties suffering from inequity. Thus such legislations as the displaced homemaker [7] are eminently desirable and to the benefit of the community. Redress is a proper domain for public policy but not necessarily attempts to preserve the marriage itself.

Efficiency considerations indicate that it is group rational for some couples to marry and then divorce, just as it is for others to marry for life and for still others not to enter in wedlock at all. Intervention on behalf of marriage can only lead to nonoptimal consequences.

§6. CHILDREN AND INTERGENERATIONAL EXTERNALITIES

The above discussion remains valid in the absence of children. Children usually have little to say in the decision to preserve or dissolve their parents' marriage. Yet they may suffer the consequences of such a decision [15,16,17 (Chap. 7)]. This constitutes an externality which may provide the basis for "pro-marriage" legislations.

An information two period life-cycle model with overlapping generations may be schematically illustrated below:

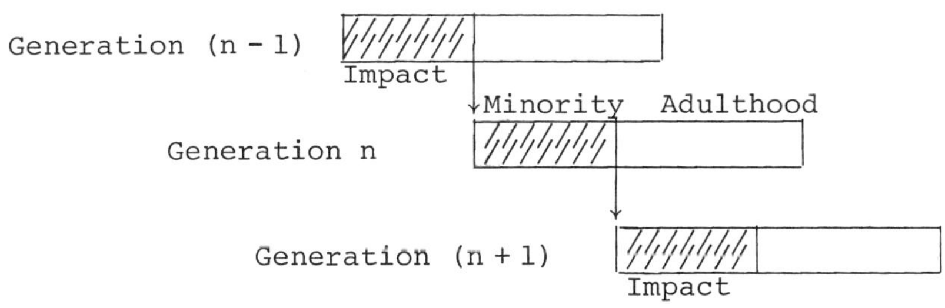

Each generation's life-cycle is represented by a bar, with the shadowed section denoting the period of minority. The two periods obviously need not be equal in duration. For simplicity, the welfare of minor children is affected by the marital status of their parents, but not by that of their grandparents.

Following the Cass and Yaari circular life-cycle model [18], consider steady-state generations chains, where the utility of generation n can be written as

$$J_n = s(t_3(n - 1)) + \int_{t_0}^{t_3(n)} u_i(c_i)dt + \int_{t_3(n)}^{T} u^i(c_i)dt \quad (27)$$

where for generation n, s is an increasing function of $t_3(n-1)$, the duration of marriage of the older generation.

An externality occurs due to the noninternalizability of the intergeneration impacts--children lack the means to "bribe" their parents to stay together. A Phelps-Polak [19] type of infinite generation game-theoretic equilibrium exists which is not optimal. Namely,

$$\underset{t_3(n)}{\text{Max}}\ J_n < \underset{t_3(n)=t_3(n-1)}{\text{Max}}\ J_n \quad (28)$$

where $\underset{t_3(n)}{\text{Max}}\ J_n$ is the Laissez Faire equilibrium payoff for each generation, not taking into account the interests of the children, while $\underset{t_3(n)=t_3(n-1)}{\text{Max}}\ J_n$ is the Pareto Optimal payoff where the intergenerational externalities are internalized through some "pro-marriage" legislation. Thus, in comparison with the Laissez Faire solution each generation would have a happier period of minority due to the fact that their parents' marriage stays intact longer, but also a less happy adulthood with enforced marital stability. The advisability of social legislation, i.e., the inequality in equation (28), implies that for each generation, the incremental happiness in the minority period outweighs the decrement of happiness in adulthood. A model with children is the subject of another paper.

ACKNOWLEDGMENT

The idea of marriage being Pareto Optimal and the impetus for modeling it owes a great deal to a discussion with Professor Murray Brown from SUNY, Buffalo.

REFERENCES

1. Nerlove, M., Toward a New Theory of Population and Economic Growth. In Economics of the Family: Marriage, Children and Human Capital, T. W. Schultz (ed.), The University of Chicago Press, 1974.
2. Schultz, T. W. (ed.), Economics of the Family: Marriage, Children and Human Capital, The University of Chicago Press, 1974.
3. Clemhout, S., and Wan, H. Y., Jr., Intra-Household Interactive Decision Making and Impact for Consumer Demand and Factory Supply (working paper), Harvard University, April 1977.
4. Clemhout, S., and Wan, H. Y., Jr., Symmetric Marriage, Household Decision Making and Impact on Fertility (working paper No. 152), Department of Economics, Cornell University, Ithaca, New York, September 1977.
5. Private Families and Public Responsibilities (mimeo), Brookings Institution, July 1977.
6. Burke, Y., H. R. 10270, 95th Congress, 1st Session.

7. Displaced Homemaker Assistance Act, Congressional Record, December 1977.

8. Kemp, M. C., and Long, N. V., Optimal Control Problems with Integrand Discontinuous with Respect to Time, The Economic Record (September 1977).

9. DaCunha, N. O., and Polak, E. (eds.), Constrained Minimization under Vector-Valued Criteria in Linear Topological Spaces. In Mathematical Theory of Control, A. V. Balakrishnan and L. W. Neustadt (eds.), Academic Press, New York, 1967.

10. Starr, A. W., and Ho, Y. C., Non-zero Sum Differential Games, Journal of Optimization Theory and Application, Vol. 3 (1969).

11. Yu, P. L., and Leitmann, G., Nondominated Decisions and Cone Convexity in Dynamic Multicriteria Decision Problems. In Multicriteria Decision Making and Differential Games, G. Leitmann (ed.), Plenum Press, New York, 1976.

12. Chakravarty, S., Optimal Savings with Finite Planning Horizon, International Economic Review, Vol. 3 (September 1962).

13. Clemhout, S., Lecture Notes on Economic Development, Department of Economics, University of California, Berkeley, California, 1966-1970.

14. Kuhn, H. W., and Tucker, A. W., Nonlinear Programming, Proceedings of the Second Berkeley Symposium on Mathematical Statistics and Probability, J. Neyman (ed.), University of California Press, Berkeley, California, 1950.

15. Wolff, S., Children Under Stress, Penguin Books, 1974.

16. Task Force on Divorce and Divorce Reform, National Council on Family Relations, Minneapolis, Minnesota, 1974.

17. Clemhout, S., Perspectives for Public Policy: An Environmental View of Human Ecology, University Press of America, Washington, D. C., 1977.

18. Cass, D., and Yaari, M. E., A Re-examination of the Pure Consumption Loans Model, Journal of Political Economy, Vol. 74, 353-67 (August 1966).

19. Phelps, E. S., and Polak, R. A., On Second Best National Survey and Game Equilibrium Growth, Review of Economic Studies, Vol. 35, 185-200 (April 1968).

A MODEL TO PREDICT THE QUALITY OF TAXI SERVICE AT A LARGE MUNICIPAL AIRPORT WHEN THE DRIVERS COMPETE

J. Case

Johns-Hopkins University
Baltimore, Maryland

In his report concerning taxi service at a certain large municipal airport, Professor DeVany constructs a model of the competition that would take place between the many cabs serving that facility, should certain rules be rescinded which presently restrict their behavior. His purpose is to show that better service can be provided at lower cost through unhindered competition than is currently available.

The *crux* of his argument is that under present conditions, many cabs return empty from the airport after dropping passengers there while others go riderless in the other direction. But very little of this so-called "deadheading" would occur under free competition. And if deadheading were reduced, the cabs would operate more efficiently, thereby permitting the reduction of fares.

The model employed is very simple. It postulates that all cabs at the airport at a given time form a single queue. So a cab driver's only options on leaving a customer at the airport are to join the queue to await a return fare, or to deadhead back to the city without one. Pure chance governs the taxi drivers' discovery of customers in the city, so the only decisions drivers make in this model are those described above, made at the airport. Thus a complete "strategy" for a driver in this "game" consists of a single number. Call it K. If he finds on arrival no more than K cabs already in the airport queue, he joins it. Otherwise he deadheads back to town.

The essential features of the model are exhibited in the following graph (Figure 1) wherein the several vertices correspond to "events in the life of a taxi and the connecting edges represent "conditions" that exist between events. One might indicate the current status of a particular cab, for instance, by a poker chip placed on the graph. If the chip is on the arc DA, the cab is en-route to the airport (a condition) with a customer. At A, the cab reaches the airport and drops its passenger (an event). Then it may choose either traveling to the city without a passenger, the condition represented by the arc AC, or waiting in the airport queue for a city-bound one, symbolized by the arc AB. The vertex B represents the event whereby such a passenger enters the cab. After it the two proceed together to a destination in the city. Vertex C represents arrival at any such destination, while D represents discovery of a new passenger in the city by the cab driver.

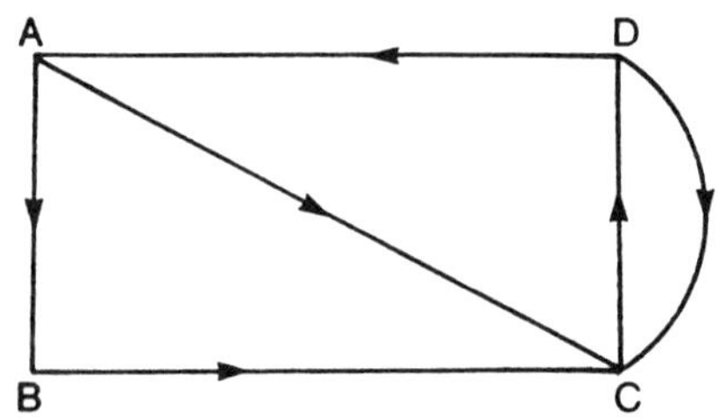

FIGURE 1

Finally, the arcs CD and DC represent, respectively, searching in the city for a customer and carrying him or her to an in-city destination.

Arcs AC, BC, and DA all require t_A hours to traverse, while DC requires t_C hours. Presumably, t_A and t_C are smaller than one. The remaining conditions (arcs) are of variable duration, because the airport queue may be either long or short, and because passengers are located in the city by a search process wherein chance is often decisive.

We shall assume that passengers turn up at the airport seeking cabs to take them to the city according to some sort of stochastic arrival process, and that their mean rate of arrival is λ passengers per hour. Similarly, passengers are generated by another such process at a mean rate of θ per hour in the city. Of these, a fraction π choose in-city destinations while the rest wish to go to the airport. Write

$$\mu = (1 - \pi)\theta \qquad \text{and} \qquad \nu = \pi\theta \tag{1}$$

If we denote by $N_{AB}(T)$ the number of complete traversals of the arc AB by taxis during the interval $0 \le t \le T$ of time, etc., we may write

$$\begin{aligned} N_{BC}(T) &= \lambda T + o(T) \quad \text{a.s.} \\ N_{DA}(T) &= \mu T + o(T) \quad \text{a.s.} \\ N_{DC}(T) &= \nu T + o(T) \quad \text{a.s.} \end{aligned} \tag{2}$$

Here, as usual, the symbol $o(T)$ denotes an otherwise unspecified function of T having the property

$$\lim_{T\to\infty} \frac{o(T)}{T} = 0 \tag{3}$$

It may be observed that

$$N_{DA}(T) = N_{AB}(T) + N_{AC}(T) + o(T) \tag{4}$$

since every cab that traverses DA during $0 \leq t \leq T$ must also traverse either AB or AC too, although possibly not until after $t = T$. In this instance, $o(T)$ stands for the number of cabs on either of arcs AB or AC when $t = T$, and cannot exceed the total number M of taxis serving the city and its environs. Clearly that number, although we do not know what number it is, satisfies condition (3).

By similarly considering the arcs incoming and outgoing at vertices B, C, and D, we may write altogether four such equations. These are

$$\begin{aligned} N_{DA}(T) &= N_{AB}(T) + N_{AC}(T) + o(T) \\ N_{AB}(T) &= N_{BC}(T) + o(T) \\ N_{AC}(T) + N_{BC}(T) + N_{DC}(T) &= N_{CD}(T) + o(T) \\ N_{CD}(T) &= N_{DA}(T) + N_{DC}(T) + o(T) \end{aligned} \tag{5}$$

The system (5) is redundant in the sense that any three of them contain all the information that the four together do, but the first, second, and fourth of them may be solved for

$$\begin{aligned} N_{AB}(T) &= \lambda T + o(T) \quad && \text{a.s.} \\ N_{AC}(T) &= (\mu - \lambda)T + o(T) \quad && \text{a.s.} \\ N_{CD}(T) &= (\mu + \nu)T + o(T) \quad && \text{a.s.} \end{aligned} \tag{6}$$

so that (2) *and* (6) describe the mean rates of taxi flow across *all* arcs of the graph.

Next, let $T_{AB}(T)$ denote the total time spent by all cabs in condition AB during $0 \leq t \leq T$, and let $T_{CD}(T)$ have like meaning for CD. Moreover, assume that of the total of M cabs available, just σM of them are in service at any given time, where $o < \sigma < 1$. Then one may write

$$[N_{AC}(T) + N_{BC}(T) + N_{DA}(T)] \cdot t_A + N_{DC}(F) \cdot t_C$$
$$+ T_{AB}(T) + T_{CD}(T) + o(T) = 2\mu T \cdot t_A + \nu T \cdot t_C$$
$$+ \frac{T_{AB}(T)}{N_{AB}(T)} N_{AB}(T) + \frac{T_{CD}(T)}{N_{CD}(T)} N_{CD}(T) + o(T) = M\sigma T \qquad (7)$$

the terms $o(T)$ being included to account for traversals not yet complete at the instant $t = T$. But the limits

$$\bar{W}_A = \lim_{T\to\infty} \frac{T_{AB}(T)}{N_{AB}(T)} \quad \text{and} \quad \bar{W}_C = \lim_{T\to\infty} \frac{T_{CD}(T)}{N_{CD}(T)} \qquad (8)$$

which are the average waiting times for the conditions AB and CD, respectively, may be shown to exist under very mild assumptions on the various passenger arrival processes. So dividing equation (7) through by T and passing to the limit as T becomes large, one obtains

$$\bar{W}_C = (\sigma M - \lambda \bar{W}_A)/\theta - 2(1 - \pi)t_A - \pi t_C \qquad (9)$$

Equation (9) expresses a necessary relationship between waiting times at the airport and in the city. It gives precise form to the intuitively obvious fact that, the more cabs choose to wait at the airport, the longer will be the queue there, and consequently the waiting time in it.

The average waiting time at the airport, $\bar{W}_A$, depends on the particular strategies adopted by the individual cab drivers. Before computing it, however, we observe that the special nature of the present graph conveys further relevant information. To this end, let us focus attention on one particular taxi cab, hereinafter called the "green cab," and let us denote by $n_{AB}(T)$ the number of complete traversals of AB by the green cab during $0 \le t \le T$, etc. Let us also write $n(T) = n_{CD}(T)$ for notational convenience.

The probability p that the green cab enters AB after DA is a function of the strategy K chosen by its driver, whom we shall call Green, and of the strategies $K_2, K_3, \ldots, K_M$ chosen by the drivers of all the other cabs. It can be zero if Green refuses ever to pick up passengers at the airport, one if he never returns to the city without a passenger, or it can assume various intermediate values. In terms of the probabilities p and π, one may write

$$\begin{aligned} n_{DC}(T) &= \pi n(T) + o(T) \\ n_{DA}(T) &= (1 - \pi)n(T) + o(T) \\ n_{AB}(T) &= p(1 - \pi)n(T) + o(T) \\ n_{AC}(T) &= (1 - p)(1 - \pi)n(T) + o(T) \\ n_{BC}(T) &= p(1 - \pi)n(T) + o(T) \end{aligned} \tag{10}$$

If Green be presumed to work throughout the period $0 \le t \le T$, he must spend all of that time in one of the six states: AB, BC, CD, DA, DC, or AC. So

$$[n_{AC}(T) + n_{BC}(T) + n_{DA}(T)] \cdot t_A + n_{DC}(T) \cdot t_C$$
$$+ t_{AB}(T) + t_{CD}(T) + o(T)$$
$$= \left[2(1-\pi)t_A + \pi t_C + \frac{t_{AB}(T)}{n_{AB}(T)} p(1-\pi) + \frac{t_{CD}(T)}{n_{CD}(T)}\right] n(T) + o(T)$$
$$= [2(1-\pi)t_A + \pi t_C + p(1-\pi)\bar{w}_A + \bar{w}_C]n(T) + o(T) = T \tag{11}$$

where $t_{AB}(T)$ and $t_{CD}(T)$ denote the number of hours spent by the green cab in states AB and CD between the times $t = 0$ and $t = T$, and

$$\bar{w}_A = \lim_{T\to\infty} \frac{t_{AB}(T)}{n_{AB}(T)} \qquad \bar{w}_C = \lim_{T\to\infty} \frac{t_{CD}(T)}{n_{CD}(T)} \tag{12}$$

Also, Green's revenues during the period would be

$$J(T) = F_A[n_{DA}(T) + n_{BC}(T)] + n_{DC}(T)F_C$$
$$= [(1-p)(1-\pi)F_A + \pi F_C]n(T) \tag{13}$$

Actually, of course, Green works only σT hours during the period $0 \le t \le T$ and earns $\sigma J(T)$ dollars. In either case, his earning rate is roughly

$$\lim_{T\to\infty} \frac{J(T)}{T} = \frac{(1+p)(1-\pi)F_A + \pi F_C}{2(1-\pi)t_A + \pi t_C + \bar{w}_C + p(1-\pi)\bar{w}_A} \text{ a.s.} \tag{14}$$

provided only that T is large. This is to be compared with the industry's earning rate of

$$\frac{(\mu + \lambda)F_A + \nu F_C}{\sigma M} \tag{15}$$

dollars per hour, the latter being the total fares collected during a typical hour, divided by the number of cab-hours worked. It is interesting to observe that if all drivers including Green adopt the same strategy, then

$$\bar{w}_A = \bar{W}_A \qquad \bar{w}_C = \bar{W}_C$$

and

$$p = \lim_{T\to\infty} \frac{n_{AB}(T)}{n_{DA}(T)} = \lim_{T\to\infty} \frac{N_{AB}(T)}{N_{DA}(T)} = \frac{\lambda}{\mu} \tag{16}$$

so the expression (14) reduces to the simpler one (15) because of (1) and (8), as plainly it must.

The symmetry of the situation suggests that, if indeed there is an "optimal strategy" for the "taxi-driver's game," it is the same for each player. So one may attempt to "solve" the game by setting the opponents' strategies K_2, K_3, ..., K_M equal to some common value N - 1 (so that Green never encounters a queue containing more than N other cabs) thereby rendering the ratio (14) dependent on just the two numbers K and N which determine p, $\bar{w}_A$, and $\bar{w}_C$, then seeking N such that K = N - 1 is an optimal strategy for Green. This task is made easier by the observation that, while Green exercises great control over his average airport waiting time $\bar{w}_A$, he is unable to materially affect his waiting times in the city, $\bar{w}_C$. Indeed $\bar{w}_C = \bar{W}_C$ if Green chooses K = N - 1, and the two differ only imperceptibly if $K \neq N - 1$. For the city is a large place, with many cabs competing for the fares generated in it, and the trouble of finding them is little altered by the duration of Green's excursions beyond. Thus one may, for practical purposes, replace $\bar{w}_C$ by $\bar{W}_C$ in (14). In view of (1) and (8), one is left finally with

$$f(K,N) = \frac{(1 + p)\mu F_A + \nu F_C}{\sigma M - \lambda \bar{W}_A + p\mu \bar{w}_A} \tag{17}$$

If Green knew N, he would surely choose K so as to maximize his long-run average earning rate f(K,N). But to determine the value of K which achieves this, one must know the exact nature of the dependence of p, $\bar{w}_A$, and $\bar{W}_A$ on K and N. And to know that, one must consider in detail the genesis of the airport queue under the assumptions that Green plays the strategy K while all the M - 1 other drivers play N - 1.

First, let us determine the probability $P_L(t)$ that the airport queue has length L at time t, Green being newly arrived at A. At such an instant, Green has long been absent from the airport, so $P_L(t)$ is virtually what it would be if K = N - 1. Hence it suffices to compute $P_L(t)$ under the additional assumption that K = N - 1. In this context, it is clear that L can have any of the values L = 0, 1, ..., N. We shall allow L to assume negative integer values as well, by adopting the convention that a queue of -L cabs means that no cabs are present and L passengers are waiting for them. Cabs and passengers can never be waiting simultaneously, as it is assumed that a passenger can enter a cab and be on his or her way to the city instantaneously. The totality of possible states of the queue is thus as indicated below in Figure 2. If the various arrival processes are in fact Poisson processes, the queue can be in

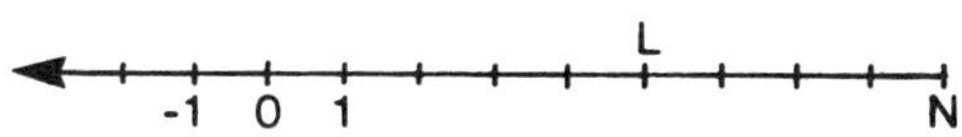

FIGURE 2

state L < N at time t + dt iff it was in one of the states L - 1, L, L + 1 at time t. Thus

$$P_L(t + dt) = P_L(t)(1 - \lambda dt)(1 - \mu dt) + P_{L-1}(t)\mu dt + P_{L+1}(t)\lambda dt \tag{18}$$

which implies

$$P_L'(t) = \mu P_{L-1}(t) - (\mu + \lambda)P_L(t) + \lambda P_{L+1}(t) \tag{19}$$

So the limits as $t \to \infty$ of the functions $P_{L-1}(\cdot)$, $P_L(\cdot)$, $P_{L+1}(\cdot)$ must be solutions of the finite difference equation

$$\alpha P_{L-1} - P_L + \beta P_{L+1} = 0 \tag{20}$$

Similarly, because the queue can be in state N at time t + dt iff it was in one of the states N or N - 1 at time t, one has

$$P_{N-1} = \rho P_N \tag{21}$$

where the constants appearing in (20) and (21) have the values

$$\alpha = \frac{\mu}{\mu + \lambda} \qquad \beta = \frac{\lambda}{\mu + \lambda} \qquad \rho = \frac{\lambda}{\mu} = \frac{\beta}{\alpha} \tag{22}$$

As is well known, the solutions of (20) are all of the form

$$P_L = A + B\rho^{-L} \qquad L < N \tag{23}$$

But since

$$\sum_{L=-\infty}^{N} P_L = 1 \tag{24}$$

A must vanish. Thus (20), (22), and (23) together imply that

$$P_L = (1 - \rho)\rho^{N-L} \tag{25}$$

for every integer L less than or equal to N. Consequently, the probability p that the queue contain no more than K cabs on Green's approach is

$$p = p(K,N) = \sum_{L=-\infty}^{K} P_L = \rho^{N-K} \tag{26}$$

Similarly Green's expected waiting time in the queue is

$$\bar{w}_A = \bar{w}_A(K,N) = \sum_{L=0}^{K} \left(\frac{L+1}{\lambda}\right) P_L = \frac{\rho^N(1-\rho)}{\lambda} \sum_{L=0}^{K} (L+1)\rho^{-L}$$

$$= \frac{\rho^N}{\lambda(1-\rho)}[(K+1)\rho^{-K} - (K+2)\rho^{1-K} + \rho^2] \tag{27}$$

while everyone else's is

$$\bar{W}_A = \bar{w}_A(N-1,N) = \frac{\rho^2}{\lambda(1-\rho)}[N/\rho - (N+1) + \rho^N] \tag{28}$$

Substitution of (26)-(28) into (14) now gives f(K,N) as an explicit function of its arguments K and N. For economy we put $K = N - J$ and write

$$f(K,N) = f(N-J,N)$$

$$= \frac{(1+\rho^J)\mu F_A + \nu F_C}{\sigma M - \rho N + (N+J-1)\rho^{2J-1} + [\rho^2(1-\rho^N) - (\rho^{2J} - \rho^{N+J-1})](1-\rho)^{-1}} \tag{29}$$

From (29) it is apparent that, unless either N or J is small, $f(\cdot,\cdot)$ is well approximated by the simpler function

$$\phi(K,N) = \phi(N-J,N) = \frac{\mu F_A + \nu F_C}{\sigma M - \rho N + \rho^2/(1-\rho)}$$

because $\rho < 1$. The significance of this fact will be discussed presently. For the moment we point only to obvious computational convenience.

The conditions that K = N - 1 be an optimal strategy for Green against his opponents' common choice of N - 1 are just

$$f(N-2,N) < f(N-1,N) > f(N,N) \tag{31}$$

which reduce to

$$\frac{2\mu F_A + \nu F_C}{\sigma M - 1/\rho - 1 - \rho + N(1/\rho - \rho) + \frac{(1/\rho - \rho^2)}{1-\rho}\rho^N} < \frac{(\mu+\lambda)F_A + \nu F_C}{\sigma M}$$

$$\frac{(1+\rho^2)\mu F_A + \nu F_C}{\sigma M + \frac{\rho^2}{1-\rho}(1+\rho-2\rho) + N\rho(\rho^2-1) + \frac{\rho}{1-\rho}(1+\rho)\rho^N} < \frac{(\mu+\lambda)F_A + \nu F_C}{\sigma M} \tag{32}$$

according to (29). It is frequently easy to solve the inequalities (32) for the values of N which satisfy them, since those N's are typically large, and the terms involving ρ^N negligible in consequence. Let us do this now for appropriate values of the several parameters indigenous to the model.

Professor DeVany's report suggests that the constants $\lambda = 35$, $\theta = 435$, $\pi = .8$, $\mu = 87$, $\nu = 348$, $\rho = .4023$, $t_A = .5$, $t_C = .2$, $F_A = 11$, $F_C = 4$, $M = 750$, and $\sigma = .6$ are reasonable descriptions of the taxi traffic between the city in question and its suburban airport. Using them (and ignoring the terms in ρ^N) reduces the inequalities (32) to

$$\frac{3306}{446.11 + 2.08N} < 6.08 > \frac{2503.87}{450.34 - .337N} \tag{32'}$$

which imply that the values of N between 47 and 114 inclusive satisfy the equilibrium conditions (31)!

So large a variety of equilibria is confusing, for classical analyses do not often deal with questions of multiplicity. Yet it is important to do so here, because the prospect of 50 cabs waiting typically at the airport is so different from that of a 110.

Figure 3 exhibits the graphs of several of the curves $y = f(K,N)$, for fixed values of N in the interval $47 \le N \le 114$. The graphs actually consist of finitely many discrete points, but these are connected with straight line segments in the figure, to illustrate certain repetitive patterns more clearly. For each N in the interval, the functions $f(K,N)$ are constant for $K > N$, and virtually so for $K < N$. Moreover, $f(K,N)$ does not exceed the industry average (15) for any K, and attains it only for $K = N - 1$. To play against a large number of opponents, all of whom behave identically, one must play exactly

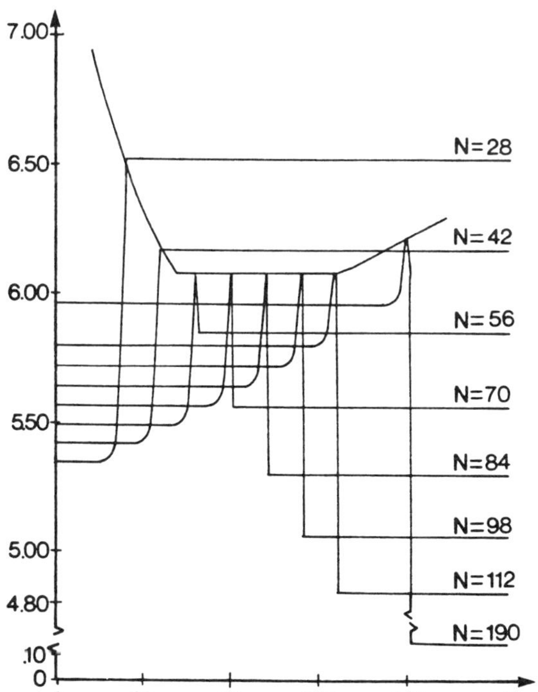

FIGURE 3

as they do. Otherwise one earns markedly less than the industry's average rate of return while the opponents earn a little more.

In practice, it is *impossible* to play exactly as one's opponents do because (a) they do *not* all behave identically, (b) one cannot measure their strategies precisely even if they are all the same, and (c) real strategies are subject to future change. It seems safe to assume, moreover, that working taxi drivers are entirely cognizant of these facts. Consequently the strategies they choose doubtless exhibit a degree of "robustness" in the sense that they yield desirable rates of return against a *considerable variety* of the opponents' possible choices, rather than being effective only against the single choice that seems most likely.

Table 1 compares the rates of return obtained from the strategies K = 12, 48, 69, 111, and 150 against various values of N with the rate

$$f^*(N) = f(K^*(N),N) \tag{33}$$

that Green would obtain if he knew his opponents' common strategy N - 1 in advance, and chose the optimal response $K = K^*(N)$ against it. It reveals, for instance, that the choice K = 69 is at least 92 percent optimal against virtually any value of N the opponents might choose. But it shows also that K = 48 is at least 90 percent optimal against those same values of N, while K = 98 is 87 percent optimal against them. And even the seemingly quite ridiculous strategies K = 12 and K = 150 are better than 72 percent optimal. So from a taxi driver's point of view, there is virtually nothing to choose between strategies

TABLE 1

N	f*(N)	f(12,N)	%	f(48,N)	%	f(69,N)	%	f(98,N)	%	f(150,N)	%
7	$7.17	$7.17	100	$7.17	100	$7.17	100	$7.17	100	$7.17	100
14	6.95	5.62	81	6.95	100	6.95	100	6.95	100	6.95	100
21	6.71	5.32	79	6.71	100	6.71	100	6.71	100	6.71	100
28	6.52	5.35	82	6.52	100	6.52	100	6.52	100	6.52	100
35	6.33	5.39	85	6.33	100	6.33	100	6.34	100	6.33	100
42	6.17	5.42	88	6.17	100	6.17	100	6.17	100	6.17	100
49	6.08	5.46	90	6.08	100	6.00	99	6.00	99	6.00	99
56	6.08	5.49	90	5.49	90	5.85	96	5.85	96	5.85	96
63	6.08	5.53	91	5.53	91	5.70	94	5.70	94	5.70	94
70	6.08	5.56	91	5.56	91	6.08	100	5.56	91	5.56	91
77	6.08	5.60	92	5.60	92	5.60	92	5.43	89	5.43	89
84	6.08	5.64	93	5.64	93	5.64	93	5.30	87	5.30	87
91	6.08	5.68	93	5.68	93	5.68	93	6.08	100	5.18	85
98	6.08	5.72	94	5.72	94	5.72	94	5.72	94	5.06	83
105	6.08	5.76	95	5.76	95	5.76	95	5.76	95	4.95	81
112	6.08	5.80	95	5.80	95	5.80	95	5.80	95	4.85	80
119	6.10	5.84	96	5.84	96	5.84	96	5.84	96	4.75	78
126	6.14	5.88	96	5.88	96	5.88	96	5.88	96	4.65	76
133	6.18	5.92	96	5.92	96	5.92	96	5.92	96	4.56	75
140	6.21	5.96	96	5.96	96	5.96	96	5.96	96	4.47	74
147	6.25	6.01	96	6.01	96	6.01	96	6.01	96	4.38	72
154	6.29	6.05	96	6.05	96	6.05	96	6.05	96	6.11	97

in the interval $48 \leq K \leq 98$, and nothing so terribly bad either about those a great deal larger or smaller. Moreover, qualitatively similar conclusions are valid if Green owns (and chooses the strategy for) not a single cab, but a small fleet of them.

From all that has now been demonstrated, the underlying nature of the competition between the taxi drivers in question is readily apparent. It is not a contest with high stakes, for strategies abound which guarantee at least 90 percent of the industry's average return. But the drivers would like to do better if they can. And since not all of them actually do behave alike, it is plain that some will earn a little more than the industry-average while their competitors get a little less.

If, for instance, a small fleet of green cabs using strategy K competes against a much larger fleet of otherwise colored ones using a common strategy $N - 1 \neq K$ in the interval $47 \leq N-1 \leq 114$, green cabs will earn less than the industry average; the others more. The driver of an independent cab belonging to neither fleet can also earn more than the industry average if he guesses which of the strategies K, $N - 1$ is the more successful, and adopts it himself.

The same is true if more than two strategies are simultaneously in use; some will be more effective than others, and there will be an incentive for the less successful players to imitate the more successful ones, until all behave and fare alike. But the strategy that emerges in this fashion seems as likely to be a matter of historical accident as to be the result of coherent economic forces. In particular, consensus seems as likely to settle on the strategy $K = 47$ as any other in the admissible interval.

For other values of the system parameters, a similar analysis is possible. In particular, an interval $N_* \le N \le N^*$ of numbers N may be found satisfying the equilibrium conditions (32), N_* being the least positive integer for which the first of those inequalities holds, and N^* being the greatest for which the other does. Almost certainly then, all cabs will join queues of length $K = N_* - 1$ or less, and the probability that there be no cabs waiting at the airport at a given instant may be estimated from (24) using $N = N_*$. That probability remains ridiculously small for a wide variety of parameter values, suggesting that taxi service at the airport will be excellent should free competition ever replace the present system at the city in question. Especially noteworthy is the fact that F_A, the fare to or from the airport, can be substantially reduced without diminishing service quality appreciably.

REFERENCE

1. DeVany, A., Alternative Ground Transportation Systems for Dallas/Fort Worth Regional Airport, a report prepared for the Federal Trade Commission, June 1977.

SOME GENERIC PROPERTIES OF COMPETITIVE EQUILIBRIUM WITH EXTERNALITIES AND WITHOUT CONVEX PREFERENCES

Kuan-Pin Lin

Department of Economics
State University of New York at Stony Brook
Stony Brook, New York

and

Harvard University
Cambridge, Massachusetts

§1. INTRODUCTION

In this paper we consider an equilibrium model with externalities, where each agent's preferences depend on the state of the economy which is specified by the allocations of each agent and a price system. This includes, as a special case, the Veblen-Scitovsky price influenced equilibrium model studied recently by Arrow and Hahn [2] and Kalman, Lin, and Wiesmeth [8]. Our analysis is restricted to pure exchange economies with a finite number of individuals and commodities.

McKenzie [10] was the first to prove explicitly the existence of equilibrium where each consumer's preference and each firm's production depend on the allocation of resources among other consumers and firms. However, this did not include Veblen-Scitovsky type economies. Arrow and Hahn [2] provide an alternative proof of existence of equilibrium for a McKenzie type model. More recently, Mas-Colell [12] and Shafer and Sonnenschein [16] prove the existence of pure exchange equilibrium with externalities without requiring complete or transitive preferences. Mantel [11] applies the same idea to a model with the presence of a complex tax structure and public goods.

It has been argued that externalities among agents in general imply the presence of nonconvexity in the associated preferences. Externalities is easy to cope with in the proof of the existence theorem while the nonconvexity is indeed an obstacle to reach a competitive equilibrium since utilities or preferences must be maximized at the equilibrium situation (see Arrow and Hahn [2] and Starrett [20]). However, it is well known that if there is a large number of agents (in Aumann's case [3], a continuum), then an appropriately defined competitive equilibrium does exist (approximately, in Starr's case [19]).

In this paper, one investigates some properties such as the local uniqueness and stability of the equilibrium for finite economies with externalities and without convex preferences. Although the equilibrium for this type of economies may fail to exist, the above properties are vacuously satisfied. Our approach, which differs from those of the above works on existence problem, uses tools of differential topology recently introduced into the economies literature by Debreu [4,5] and Smale [17,18]. In particular, applying Thom's transversality theory of differential topology, the above equilibrium properties are generic. That is, the equilibrium is locally unique and stable for "almost all" economies. Recently, Fuchs and Laroque [6] obtained local uniqueness and stability of equilibria for a McKenzie-type model using a demand function approach. We study utility functions directly and do not require well defined demand functions. In a similar direction, using a distribution approach, Mas-Colell [13] established the same generic properties of competitive equilibrium for large economies with a stronger requirement in defining the concept of "almost all" economies. By allowing externalities and nonconvex preferences together, our results extend Debreu [4] and Smale [17] works on the comparative statics of equilibria for finite economies.

§2. THE MODEL

We consider a space of economies with ℓ commodities and n agents. Let $P = \{z \in R^{\ell} : z_i > 0, i = 1, \ldots, \ell\}$ be the consumption space for agent h (h = 1, ..., n), and $S = \{z \in P : \sum_{i=1}^{\ell} z_i = 1\}$ be the price space. A *state* of an economy is a pair $(x,p) \in P^n \times S$ where $x = (x^1, \ldots, x^n)$. We denote the product space $P^n \times S$ as the state space. In economies with externalities, the preferences of agent h can be represented by real-valued functions defined on the state space, i.e., $u^h : p^n \times S \to R$ (see [8] and [16], for example). In other words, we allow the preferences of each agent to depend not only on his own consumption, but also on the consumption of others and the price system. It is understood that u^h represents a preference ordering over the state space, which is not necessarily convex. For each agent h, we assume $u^h \in C^2(p^n \times S,R)$[1] and $D_h u^h(x,p) \in P$ for each $(x,p) \in p^n \times S$, where

---

[1]In general, $C^k(X,Y)$ denotes the space of k times continuously differentiable functions from X to Y.

$D_h u^h(x,p)$ is the derivative of u^h with respect to the hth coordinate of vector x. The latter is simply a monotonicity assumption: each commodity is desired by each agent. For each agent h, let $\mathcal{U}$ be the space of utility functions satisfying the above assumptions. That is

$$\mathcal{U} = \{u^h \in C^2(P^n \times S,R) : D_h u^h(x,p) \in P \text{ for each } (x,p) \in P^n \times S\}$$

Furthermore, we let $\bar{x}^h \in P$ be the resource endowment for the agent h. Since the endowment as well as its distribution of each agent are allowed to vary in the space P, the economic characteristics of our model are completely specified by the product space $(\mathcal{U} \times P)^n$. Let $\mathcal{E} = (\mathcal{U} \times P)^n$ be the space of economies, and an economy with externalities and without convex preferences is a list of utility functions in $\mathcal{U}^n$ and resource endowments in P^n, i.e., $E = (u,\bar{x})$ with $u = (u^1,\ldots,u^n)$ and $\bar{x} = (\bar{x}^1,\ldots,\bar{x}^n)$. Clearly, $\mathcal{E}$ is an infinite dimensional space. For dealing with "generic" properties of competitive equilibrium, $\mathcal{E}$ will be endowed with a topology which is defined by the product of the induced Whitney C^2 topology[2] on $\mathcal{U}^n$ and the induced usual topology on P^n. It is an extremely large topology which controls the closeness of the maps and their derivatives.

---

[2]A neighborhood basis of $f \in C^k(R^m,R)$ in the Whitney C^k topology is given by the sets $N^\delta(f)$, with $\delta : R^m \to R$ positive continuous and $N^\delta(f) = \{f' \in C^k(R^m,R) : \|D^\alpha f'(x) - D^\alpha f(x)\| < \delta(x)$ for all $x \in R^m$ and $\alpha = 0,1,\ldots,k\}$ (see [7] or [17] for details).

Now, we are in a position to define two concepts of equilibrium. For every economy $E \in \mathcal{E}$, a competitive equilibrium is a state (x,p) with $\sum_{h=1}^{n} x^h = \sum_{h=1}^{n} \bar{x}^h$ and x^h is a maximal point of $u^h(x^1,\ldots,x^{h-1},.,x^{h+1},\ldots,x^n,p)$ satisfying $px^h = p\bar{x}^h$ for every h. Formally, we define the set of competitive equilibria for $E \in \mathcal{E}$ as

$$W(E) = \Big\{(x,p) \in P^n \times S : u^h(x,p) \text{ is maximized,}$$

$$px^h = p\bar{x}^h,\ h = 1,\ldots,n, \text{ and } \sum_{h=1}^{n} x^h = \sum_{h=1}^{n} \bar{x}^h\Big\}$$

We recall that if $f : X \to Y$ is class C^1, a point $x \in X$ is a regular point of f if $Df(x) : T_xX \to T_yY$ is surjective with $y = f(x)$ where $Df(x)$ represents the derivative of the map f computed at x, which is a linear map from the tangent space of X at x to the tangent space of Y at y, denoted by T_xX and T_yY, respectively. If $Df(x)$ is not surjective, x is a critical point of f. y is called a regular value if every $x \in f^{-1}(y)$ is a regular point. y is a critical value if at least one $x \in f^{-1}(y)$ is a critical point. Clearly, if x^h is a maximal point of $u^h(x^1,\ldots,x^{h-1},.,x^{h+1},\ldots,x^n,p)$ subject to $px^h = p\bar{x}^h$, x^h is also a critical point of it. For a given parameter vector $(x^1,\ldots,x^{h-1},x^{h+1},\ldots,x^n,p)$, the condition for x^h to be a critical point of $u^h(x^1,\ldots,x^{h-1},.,x^{h+1},\ldots,x^n,p)$ subject to $px^h = p\bar{x}^h$ can be written as $D_hu^h(x,p) = \lambda^h p$, where λ^h is the Lagrangian multiplier of h. To avoid λ^h in the model, we substitute

$$\lambda^h = |D_hu^h(x,p)|$$

where

$$|D_h u^h(x,p)| = \sum_{i=1}^{\ell} \frac{\partial u^h}{\partial x_i^h}(x,p)$$

It is obvious that $\lambda^h > 0$ by monotonicity of u^h. Hence the set of extended competitive equilibria for $E \in \mathcal{E}$ is as follows:

$$\Phi(E) = \Big\{(x,p) \in P^n \times S : D_h u^h(x,p) = |D_h u^h(s,p)|\cdot p,$$
$$px^h = p\bar{x}^h,\ h = 1,\ldots,n, \text{ and } \sum_{h=1}^{n} x^h = \sum_{h=1}^{n} \bar{x}^h\Big\}$$

The concept of extended equilibria for the economies without external effects was first used by Smale [17]. Since the condition $px^n = p\bar{x}^n$ can be obtained from $px^h = p\bar{x}^h$, $h = 1,\ldots,n-1$, and $\sum_{h=1}^{n} x^h = \sum_{h=1}^{n} \bar{x}^h$, the set of extended equilibria for every $E \in \mathcal{E}$ can be rewritten as

$$\Phi(E) = \Big\{(x,p) \in P^n \times S : D_h u^h(x,p) = |D_h u^h(x,p)|\cdot p,$$
$$px^h = p\bar{x}^h,\ h = 1,\ldots,n-1, \text{ and } \sum_{h=1}^{n} x^h = \sum_{h=1}^{n} \bar{x}^h\Big\}$$

For every economy $E = (u,\bar{x}) \in \mathcal{E}$, we define a map $\psi_E : P^n \times S \to I^n \times R^{n+\ell-1}$ by

$$\psi_E(x,p) = \Big\{D_h u^h(x,p) - |D_h u^h(x,p)|\cdot p,\ h = 1,\ldots,n;$$
$$p\bar{x}^h - px^h,\ h = 1,\ldots,n-1;\ \sum_{h=1}^{n} \bar{x}^h - \sum_{h=1}^{n} x^h\Big\}$$

where

$$I = \Big\{z \in R^\ell : \sum_{i=1}^{\ell} z_i = 0\Big\} \subset R^\ell$$

Obviously, $\psi_E \in C^1(P^n \times S,\ I^n \times R^{n+\ell-1})$. By definition of

$\Phi(E)$, we have $\Phi(E) = \psi_E^{-1}(0)$ and $W(E) \subset \psi_E^{-1}(0)$ for every $E \in \mathcal{E}$. That is, if (x,p) is a competitive equilibrium, it is an extended equilibrium, and the C^1 map ψ_E vanishes at (x,p). Furthermore, $\Phi(E)$ is closed in $P^n \times S$ since $\Phi(E) = \psi_E^{-1}(0)$ and ψ_E is C^1.

§3. LOCAL UNIQUENESS AND STABILITY OF EQUILIBRIA

In this section we study local uniqueness and stability of competitive equilibria for "almost all" economies in $\mathcal{E}$, which is defined by a regularity condition on ψ_E below. We first recall a concept of transversality in differential topology (see [1]). Let Z be a submanifold of Y, $f \in C^1(X,Y)$ is said to be *transversal* to Z at x, denoted by $f \pitchfork_x Z$, if either $y = f(x) \notin Z$ or $y = f(x) \in Z$ and $Df(x)[T_xX] + T_yZ = T_yY$, i.e., the image of $Df(x)$ and T_yZ as subspaces of T_yY span T_yY. If $f \pitchfork_x Z$, for every $x \in X$, $f \pitchfork Z$. In view of the space of economies, the notion of transversality makes sense in the model. We claim that the subspace of economies satisfying a transversality requirement is open and dense. Actually, we apply the concept of transversality only in the very special sense of the above consideration. That is, Z is just a single point y, and therefore its tangent space is the zero subspace of T_yY. Thus, f is transversal to y if either $y \neq f(x)$ for all x or $Df(x)[T_xX] = T_yY$ for all $x \in f^{-1}(y)$, which is to say that y is a regular value of f. So transversality includes the notion of regularity as a special case. In fact, f is regular if and only if $f \pitchfork y$ for every $y \in Y$. We need a few more definitions. An element $E \in \mathcal{E}$ is called a *regular economy* if and only if the associated map ψ_E is transversal to the origin, i.e.,

$\psi_E \pitchfork 0$. Moreover, the *space of regular economies* is denoted by the set $\mathcal{R} = \{E \in \mathcal{E} : \psi_E \pitchfork 0\}$. By a theorem of differential topology (for instance, [1], p. 45), $\psi_E^{-1}(0) = \Phi(E)$ is a C^1 submanifold of $P^n \times S$ for every $E \in \mathcal{R}$. In view of the space of economies $\mathcal{E}$, we have

Proposition 1 *$\mathcal{R}$ is open and dense*[3] *in $\mathcal{E}$.*

Proof See Section 4.

For every regular economy $E \in \mathcal{R}$, we have $\psi_E \pitchfork 0$. By the openness property of $\mathcal{R}$, $\psi_{E'} \pitchfork 0$ for $E' \in \mathcal{R}$ sufficiently near E. One might expect that for E' near E, $\psi_{E'}^{-1}(0)$ and $\psi_E^{-1}(0)$ are close to each other. In other words, we have

Theorem 1 *The extended equilibrium correspondence Φ defined by $\Phi(E) = \psi_E^{-1}(0)$ for $E \in \mathcal{R}$ is continuous, i.e., it is stable for $E \in \mathcal{R}$.*

Proof See Section 4.

Corollary 1 *The competitive equilibrium $W(E)$ is stable for $E \in \mathcal{R}$.*

Next, since $\Phi(E) = \psi_E^{-1}(0)$ is closed for every $E \in \mathcal{E}$ and $\psi_E^{-1}(0)$ is a submanifold with zero dimension if $E \in \mathcal{R}$, thus $\Phi(E)$ is a discrete set. Hence, we have

Theorem 2 *For every regular economy $E = (u,\bar{x}) \in \mathcal{R}$, the extended equilibrium set $\Phi(E)$ is locally unique.*

---

[3]The density and openness of $\mathcal{R}$ in $\mathcal{E}$ implies that any economy can be approximated by a regular economy and any regular economy is still regular under small perturbations of economic data in the model.

Corollary 2 *For every regular economy* $E \in \mathcal{R}$, *the competitive equilibrium set* $W(E)$ *is locally unique.*

§4. PROOFS

Proof of Proposition 1 Since P and S are locally compact, we let $\{K_\alpha\}$ and $\{L_\alpha\}$ be the sequences of compact subsets in P and S, respectively, such that $K_\alpha \subset \mathring{K}_{\alpha+1}$, $L_\alpha \subset \mathring{L}_{\alpha+1}$ and $P = \cup K_\alpha$, $S = \cup L_\alpha$. For each $u^h \in C^2(P^n \times S, R)$, let

$$u_\alpha^h = u^h | K_\alpha^n \times L_\alpha \in C^2(K_\alpha^n \times L_\alpha, R)$$

The spaces $C^2(K_\alpha^n \times L_\alpha, R)$ are Banach spaces (see [1], p. 24). Moreover, they are metrizable and separable, hence they are second countable. It is easy to see that the space $C^1(P^n \times S, R)$ is the inverse limit[4] of the sequence $\{C^2(K_\alpha^n \times L_\alpha, R), f_\alpha\}$. That is,

$$f_\alpha : C^2(K_\alpha^n \times L_\alpha, R) \to C^2(K_{\alpha-1}^n \times L_{\alpha-1}, R)$$

defined by $f_\alpha(u_\alpha^h) = u_{\alpha-1}^h = u_\alpha^h | K_{\alpha-1}^n \times L_{\alpha-1}$ is clearly continuous. Define

$$\mathcal{U}_\alpha = \{u_\alpha^h \in C^2(K_\alpha^n \times L_\alpha, R) : D_h u_\alpha^h(x,p) \in P$$
$$\text{for each } (x,p) \in K_\alpha^n \times L_\alpha\}$$

and $\mathcal{E}_\alpha = (\mathcal{U}_\alpha \times P)^n$. Then $\mathcal{U}$ and $\mathcal{E}$ are the inverse limit

---

[4]Let X_α be a topological space and f_α be a continuous map from X_α into $X_{\alpha-1}$, for each α. The sequence $\{X_\alpha, f_\alpha\}$ is called an inverse limit sequence. The inverse limit space of the sequence $\{X_\alpha, f_\alpha\}$ is the following subset of $\Pi_\alpha X_\alpha : X = \{x \in \Pi_\alpha X_\alpha :$ $f_\alpha(x_\alpha) = x_{\alpha-1}$ for each α and $x_\alpha \in X_\alpha$, $x_{\alpha-1} \in X_{\alpha-1}\}$ (see [22] for its formal definition and properties).

spaces of the sequences $\{\mathcal{U}_\alpha, f'_\alpha\}$ and $\{\mathcal{E}_\alpha, g_\alpha\}$, respectively, where $f'_\alpha = f_\alpha | \mathcal{U}_\alpha$, $g_\alpha : \mathcal{E}_\alpha \to \mathcal{E}_{\alpha-1}$ defined by

$$g_\alpha = (\underbrace{f'_\alpha, \ldots, f'_\alpha}_{n \text{ times}}, \text{id})$$

and id, the identity map, from P^n to P^n. Clearly, $\mathcal{E}_\alpha$ is a C^2 (Banach) manifold and second countable. Define the sequence $\{\mathcal{R}_\alpha, g'_\alpha\}$ as

$$\mathcal{R}_\alpha = \{E_\alpha \in \mathcal{E}_\alpha : \psi_{E_\alpha} \pitchfork 0\}, \quad g'_\alpha = g_\alpha | \mathcal{R}_\alpha$$

and

$$\psi_{E_\alpha} = \psi_E | K^n_\alpha \times L_\alpha$$

Then $\mathcal{R}$ is the inverse limit of $\{\mathcal{R}_\alpha, g'_\alpha\}$.

Observe that although $K^n_\alpha \times L_\alpha$ is not a manifold in the usual sense, we can define the tangent space of $K^n_\alpha \times L_\alpha$ in a boundary point unambiguously: Let $\theta \subset R^{\ell n+\ell-1}$ be an open neighborhood of $K^n_\alpha \times L_\alpha$. For $(x,p) \in K^n_\alpha \times L_\alpha$, define $T_{(x,p)} K^n_\alpha \times L_\alpha = T_{(x,p)}\theta$. By this and the differentiability on the boundary of $K^n_\alpha \times L_\alpha$, it follows that ψ_{E_α} is C^1 in the above sense in a neighborhood of $K^n_\alpha \times L_\alpha$, and that the notion of transversality makes sense in our context. Hence, the set $\mathcal{R}_\alpha$ is meaningful.

We now claim that $\mathcal{R}_\alpha$ is an open and dense in $\mathcal{E}_\alpha$ for each α. We apply the Transversal Density Theorem 19.1 of [1], p. 48. Conditions (1), (2), and (3) of 19.1 are satisfied. We need to check condition (4) of Theorem 19.1 of [1]. First, let $\psi_\alpha : \mathcal{E}_\alpha \times K^n_\alpha \times L_\alpha \to I^n \times R^{n+\ell-1}$ defined by $\psi_\alpha(E_\alpha, x, p) = \psi_E(x,p)$ for each $E_\alpha \in \mathcal{E}_\alpha$ and $(x,p) \in K^n_\alpha \times L_\alpha$ be the evaluation map of ψ_{E_α}. It is clear that ψ_α is C^1 (for instance, see [1], p. 25). We go on to prove that $\psi_\alpha \pitchfork 0$. By definition, $\psi_\alpha : \mathcal{E}_\alpha \times K_\alpha^n \times L_\alpha \to I^n \times R^{n+\ell-1}$ is given by

$$\psi_\alpha(E_\alpha,x,p) = \Big\{ D_h u_\alpha^h(x,p) - |D_h u_\alpha^h(x,p)|\cdot p,\ h = 1,\ldots,n;$$

$$p\bar{x}^h - px^h,\ h = 1,\ldots,n-1;\ \sum_{h=1}^n \bar{x}^h - \sum_{h=1}^n x^h \Big\}$$

Its derivative

$$D\psi_\alpha(E_\alpha,x,p) : T_{(E_\alpha,x,p)}(\mathcal{E}_\alpha \times K_\alpha^n \times L_\alpha)$$

$$\to T_{\psi_\alpha(E_\alpha,x,p)}(I^n \times R^{n+\ell-1})$$

at (E_α,x,p) is defined by

$$D\psi_\alpha(E_\alpha,x,p)(\dot{E}_\alpha,\dot{x},\dot{p}) = \Bigg\{ \frac{\partial^2 u_\alpha^h}{\partial x_r^h \partial E_\alpha} \dot{E}_\alpha - \sum_{i=1}^\ell \frac{\partial^2 u_\alpha^h}{\partial x_i^h \partial E_\alpha} p_r \dot{E}_\alpha$$

$$+ \sum_{k=1}^n \sum_{j=1}^\ell \frac{\partial^2 u_\alpha^h}{\partial x_r^h \partial x_j^k} \dot{x}_j^k - \sum_{k=1}^n \sum_{j=1}^\ell \sum_{i=1}^\ell \frac{\partial^2 u_\alpha^h}{\partial x_i^h \partial x_j^k} p_r \dot{x}_j^k$$

$$+ \sum_{j=1}^\ell \frac{\partial^2 u_\alpha^h}{\partial x_r^h \partial p_j} \dot{p}_j - \sum_{j=1}^\ell \sum_{i=1}^\ell \frac{\partial^2 u_\alpha^h}{\partial x_i^h \partial p_j} p_r \dot{p}_j$$

$$- \sum_{i=1}^\ell \frac{\partial u_\alpha^h}{\partial x_i^h} \dot{p}_r,\quad r = 1,\ldots\ell;\ h = 1,\ldots,n$$

$$\dot{p}(\bar{x}^h - x^h) + p(\dot{\bar{x}}^h - \dot{x}^h),\ h = 1,\ldots,n-1;$$

$$\sum_{h=1}^n \dot{\bar{x}}^h - \sum_{h=1}^n \dot{x}^h \Bigg\}$$

where

$$(\dot{E}_\alpha,\dot{x},\dot{p}) \in T_{(E_\alpha,x,p)}(\mathcal{E}_\alpha \times K_\alpha^n \times L_\alpha)$$

and $\dot{E}_\alpha = (\dot{u}_\alpha, \dot{\bar{x}})$. Without loss of generality, we take $\dot{E}_\alpha = (0, \dot{\bar{x}})$ and $\dot{x} = 0$, then

$$D\psi_\alpha(E_\alpha, x, p)((0,\dot{\bar{x}}), 0, \dot{p})$$

$$= \left\{ \sum_{j=1}^{\ell} \frac{\partial^2 u_\alpha^h}{\partial x_r^h \partial p_j} \dot{p}_j - \sum_{j=1}^{\ell} \sum_{i=1}^{\ell} \frac{\partial^2 u_\alpha^h}{\partial x_i^h \partial p_j} p_r \dot{p}_j - \sum_{i=1}^{\ell} \frac{\partial u_\alpha^h}{\partial x_i^h} \dot{p}_r, \right.$$

$$r = 1, \ldots, \ell,\ h = 1, \ldots, n;$$

$$\left. \dot{p}(\bar{x}^h - x^h) + p\dot{\bar{x}}^h,\ h = 1, \ldots, n-1;\ \sum_{h=1}^{n} \bar{x}^h \right\}$$

For each

$$(a,b,c) \in T_{\psi_\alpha(E_\alpha,x,p)}(I^n \times R^{n+\ell-1})$$

with $a = (a^1, \ldots, a^n) \in T_{\psi_\alpha(E_\alpha,x,p)}(I^n)$, $b = (b^1, \ldots, b^{n-1}) \in R^{n-1}$ and $c = (c_1, \ldots, c_\ell) \in R^\ell$, there exists

$$((0,\dot{\bar{x}}), 0, \dot{p}) \in T_{(E_\alpha,x,p)}(\mathcal{E}_\alpha \times K_\alpha^n \times L_\alpha)$$

such that $D\psi_\alpha(E_\alpha,x,p)((0,\dot{\bar{x}})0,\dot{p}) = (a,b,c)$ since

$$\sum_{k=1}^{\ell} a_k^h = 0 \qquad \sum_{i=1}^{\ell} \frac{\partial u_\alpha^h}{\partial x_i^h} > 0$$

for each h and $p \neq 0$. Therefore $D\psi_\alpha(E_\alpha,x,p)$ is surjective on $T_{\psi_\alpha(E_\alpha,x,p)}(I^n \times R^{n+\ell-1})$. In particular, $\psi \pitchfork 0$. This shows that condition (4) of the Transversal Density Theorem 19.1 of [1] is satisfied. Hence $\mathcal{R}_\alpha$ is dense in $\mathcal{E}_\alpha$. The openness of $\mathcal{R}_\alpha$ in $\mathcal{E}_\alpha$ follows from the Openness of Transversal Intersection Theorem 18.2 of [1], p. 47.

It remains to prove that $\mathcal{R}$ is open and dense in $\mathcal{E}$ with respect to the topology constructed in Section 2. Let $r_\alpha : \mathcal{E} \to \mathcal{E}_\alpha$ for every α be the canonical restriction maps.

We first claim that $r_\alpha^{-1}(\mathcal{R}_\alpha)$ is dense in $\mathcal{E}$. We note that r_α is not an open map with respect to the topology on $\mathcal{E}$. But in fact, we do not need the openness of r_α, and it would suffice if we know that the image of an open set of $\mathcal{E}$ under r_α contains an open set of $\mathcal{E}_\alpha$. Let $N(E) = N^\varepsilon(u) \times N(\bar{x})$ be a neighborhood of $E = (u,\bar{x})$ in $\mathcal{E}$, where $N(\bar{x})$ is a usual neighborhood of $\bar{x}$ in P and $N^\varepsilon(u) = \{u' \in \mathcal{U}^n : \|D^k u^h(x,p) - D^k u^{h'}(x,p)\| < \varepsilon^h(x,p)$ for all $(x,p) \in P^n \times S$, $k = 0,1,2$ and $h = 1,\ldots,n\}$ with $\varepsilon^h : P^n \times S \to R$ being a positive continuous function for each h. As we discuss earlier, $r_\alpha(N(E)) \subset \mathcal{E}_\alpha$ is not an open set in general. However, if we shrink $N(E)$ to a neighborhood $N^*(E) = N^\delta(u) \times N(\bar{x})$ with $\delta^h \leq \varepsilon^h$ and $\delta^h : P^n \times S \to R$ is a positive continuous function and increasing with respect to $x^h \in P$ for every h, it is obvious that for every $E'_\alpha = (u'_\alpha, \bar{x}') \in N^*_\alpha(E_\alpha) = N^\delta_\alpha(u_\alpha) \times N(\bar{x})$, $u^{h'}_\alpha$ can be extended to a function $u^{h'} \in \mathcal{U}$ with $u^{h'}_\alpha = u^{h'}|K^n_\alpha \times L_\alpha$ for every h, where $N^\delta_\alpha(u_\alpha) = \{u'_\alpha \in \mathcal{U}^n_\alpha :$ $\|D^k u^h_\alpha(x,p) - D^k u^{h'}_\alpha(x,p)\| < \delta^h(x,p)$ for every $(x,p) \in K^n_\alpha \times L_\alpha$, $k = 0,1,2$ and $h=1,\ldots,n\}$. Hence $r_\alpha(N^*(E)) = N^*_\alpha(E_\alpha)$ is open in $\mathcal{E}_\alpha$ and consequently $r_\alpha(N(E))$ contains an open set. Together with the fact that $\mathcal{R}_\alpha$ is dense in $\mathcal{E}_\alpha$, we have $r_\alpha(N(E)) \cap \mathcal{R}_\alpha \neq \phi$. This means that there exists an $E' \in N(E)$ such that $r_\alpha(E') \in \mathcal{R}_\alpha$ or $E' \in r_\alpha^{-1}(\mathcal{R}_\alpha)$. Hence $N(E) \cap r_\alpha^{-1}(\mathcal{R}_\alpha) \neq \phi$, or equivalently, $r_\alpha^{-1}(\mathcal{R}_\alpha)$ is dense in $\mathcal{E}$. By definition, $\mathcal{R} = \cap_\alpha r_\alpha^{-1}(\mathcal{R}_\alpha)$. Therefore, $\mathcal{R}$ is dense in $\mathcal{E}$ since $\mathcal{E}$ is a Baire space. Moreover, if $E \in \mathcal{R}$, then by definition, $E_\alpha \in \mathcal{R}_\alpha$ with $g'_\alpha(E_\alpha) = E_{\alpha-1}$ for each α. Since $\mathcal{R}_\alpha$ is open in $\mathcal{E}_\alpha$, there exists a neighborhood $N_\alpha(E_\alpha) = N^\varepsilon_\alpha(u_\alpha) \times N(\bar{x})$ of E_α in $\mathcal{E}_\alpha$ with $N_\alpha(E_\alpha) \subset \mathcal{R}_\alpha$ for each α. In particular,

$$N_\alpha^\varepsilon(u_\alpha) = \{u'_\alpha \in \mathcal{U}_\alpha^n : \|D^k u_\alpha^h(x,p) - D^k u_\alpha^{h'}(x,p)\| < \varepsilon^h(x,p)$$

for every $(x,p) \in K_\alpha^n \times L_\alpha$, $k = 0,1,2$ and $h = 1,\ldots,n\}$, where $\varepsilon^h : K_\alpha^n \times L_\alpha \to R$ is a positive continuous function for every h. We now choose a positive continuous function $\delta^h : P^n \times S \to R$ with $\delta^h(x,p) \le \varepsilon^h(x,p)$ for every $(x,p) \in K_\alpha^n \times L_\alpha$ and all α. Then $N^*(E) = N^\delta(u) \times N(\bar{x})$ is a neighborhood of E in $\mathcal{E}$ and $N^*(E) \subset \mathcal{R}$. Hence the openness of $\mathcal{R}$ in $\mathcal{E}$ follows. This completes the proof.

<u>Proof of Theorem 1</u> We know that ψ_α is C^1. Moreover, for every $E_\alpha \in \mathcal{R}_\alpha$, ψ_{E_α} is a C^1 local diffeomorphism by the inverse function theorem since

$$D\psi_{E_\alpha}(x,p) : T_{(x,p)}(K_\alpha^n \times L_\alpha) \to T_{\psi_{E_\alpha}(x,p)}(I^n \times R^{n+\ell-1})$$

with $(x,p) \in \psi_E^{-1}(0)$ is an isomorphism (see [14]). Hence, the stability property of the map $\Phi_\alpha = \Phi | \mathcal{R}_\alpha \to P^n \times S$ follows from an application of the implicit function theorem on the evaluation map ψ_α. That is, there exist neighborhoods $N_\alpha(E_\alpha)$ of $E_\alpha \in \mathcal{R}_\alpha$ and V of $(x,p) \in K_\alpha^n \times L_\alpha \subset P^n \times S$, and a C^1 function $\xi_\alpha : N_\alpha(E_\alpha) \to V$ such that $\psi_\alpha(E'_\alpha, \xi_\alpha(E'_\alpha)) = 0$ for every $E'_\alpha \in N_\alpha(E_\alpha)$ and $\xi_\alpha(E_\alpha) = (x,p)$. Since $\Phi_{\alpha-1}(E_{\alpha-1}) \subset \Phi_\alpha(E_\alpha)$ for every α, we have the following diagram

$$\begin{array}{ccc}
N_{\alpha-1}(E_{\alpha-1}) & \xleftarrow{\;g'_\alpha | N_\alpha(E_\alpha)\;} & N_\alpha(E_\alpha) \\
\downarrow \xi_{\alpha-1} & & \downarrow \xi_\alpha \\
V & \xleftarrow{\;\text{id}\;} & V
\end{array}$$

which is commutative, i.e., $\xi_{\alpha-1} \circ g' | N_\alpha(E_\alpha) = \text{id} \circ \xi_\alpha$ for every α. This implies that for each $E \in \mathcal{R}$ there is a continuous function $\xi : N^*(E) \to V$ such that $\psi(E',\xi(E')) = 0$ for every $E' \in N^*(E)$ and $\xi(E) = (x,p)$, where $N^*(E)$ is a neighborhood as described in the proof of Proposition 1. Hence the extended equilibrium correspondence Φ is stable for $E \in \mathcal{R}$. This completes the proof.

REFERENCES

1. Abraham, R., and Robbin, J., Transversal Mappings and Flows, W. A. Benjamin, Inc., New York, 1967.
2. Arrow, K. J., and Hahn, F. H., General Competitive Analysis, Holden-Day, San Francisco, 1971.
3. Aumann, R., Existence of Competitive Equilibria in Markets with a Continuum of Traders, Econometrica 34, 1-17 (1963).
4. Debreau, G., Economies with a Finite Set of Equilibria, Econometrica 38, 387-392 (1970).
5. Debreau, G., Smooth Preferences, Econometrica 40, 603-615 (1972).
6. Fuchs, G., and Laroque, G., Continuity of Equilibria for Economies with Vanishing External Effects, Journal of Economic Theory, 9, 1-22 (1974).
7. Hirsh, M. W., Differential Topology, Springer Verlag, New York, 1976.
8. Kalman, P. J., Lin, K.-P., and Wiesmeth, H., Equilibrium Theory in Veblen-Scitovsky Economies: Local Uniqueness Stability and Existence, Working Paper, Harvard University, November, 1975.
9. Kelley, J., General Topology, Van Nostrand Co., Princeton, N. J., 1960.
10. McKenzie, L. W., Competitive Equilibrium with Dependent Consumer Preferences, in National Bureau of Standards

and Department of the Air Force, The Second Symposium on Linear Programming, Washington, D. C., 1955.

11. Mantel, R. R., General Equilibrium and Optimal Taxes, Journal of Mathematical Economics, 2, 187-200 (1975).

12. Mas-Colell, A., An Equilibrium Existence Theorem without Complete or Transitive Preferences, Journal of Mathematical Economics, 1, 237-246 (1974).

13. Mas-Colell, A., Regular Nonconvex Economies, Econometrica, 45, 1387-1407 (1977).

14. Milnor, J., Topology from the Differentiable Viewpoint, University of Virginia Press, 1965.

15. Scitovsky, T., Some Consequences of Habit of Judging Quality by Price, Review of Economic Studies, 12, 100-105 (1945).

16. Shafer, W., and Sonnenschein, H., Some Theorems on the Existence of Competitive Equilibrium, Journal of Economic Theory, 11, 83-93 (1975).

17. Smale, S., Global Analysis and Economics IIA: Existence of a Theorem of Debreu, Journal of Mathematical Economics, 1, 1-14 (1974).

18. Smale, S., Global Analysis and Economics IV: Finiteness and Stability of Equilibria with General Consumption Sets and Production, Journal of Mathematical Economics, 1, 119-127 (1974).

19. Starr, R., Quasi-equilibrium in Markets and Non-convex Preferences, Econometrica, 37, 25-38 (1969).

20. Starrett, D., Fundamental Nonconvexities in the Theory of Externalities, Journal of Economic Theory, 4, 180-199 (1972).

21. Veblen, T., The Theory of the Leisure Class, Macmillan, New York, 1899.

22. Wallace, A. H., Algebraic Topology: Homology and Cohomology, Benjamin, New York, 1970.

BAYESIAN CONVERGENCE IN MARKET GAMES WITH INCOMPLETE INFORMATION

H. M. Shefrin

Department of Economics, University of Rochester
Rochester, New York

and

Department of Economics, University of Santa Clara
Santa Clara, California

This paper describes a new Bayesian treatment of games with incomplete information, which is useful in studying imperfect economic markets. The approach is also related to optimal control problems that can be formulated as Markov decision processes.

In the real world markets are highly complex institutions. There are a great many participants, market information is diversely distributed, and decisions are arrived at in the face of imperfect knowledge. Significantly, an individual participant's knowledge about the plans of the other participants is usually incomplete. Traditionally, game theoretic treatments of markets have ignored the imperfections associated with incomplete information. In order to explain the contribution of the present paper, it is useful to describe some of the past approaches.

Standard competitive market game models assume the existence of an auctioneer. This auctioneer is solely responsible for the setting of all prices. Each market participant responds to these prices by attempting to implement its best 'affordable' trade plan. Given these plans, the auctioneer attempts to choose prices in order to maximize the value of excess demand. It turns out that the game has a Nash equilibrium and that this equilibrium has a very attractive property: demand and supply are equal for all commodities. Furthermore, the Nash equilibrium belongs to the core of this game, and with 'enough' market participants, is the only member of the core.

I would contend that the above approach is not appropriate for the analysis of highly complex economic systems. In the real world, it is rare to find an auctioneer: market participants themselves set prices. Traders interact with other traders directly, not through a fictitious intermediary. Consequently, information about planned trades and market prices is likely to be diverse, imperfect, and variable over time.

Harsanyi [2] draws an important distinction between games with imperfect information and games with incomplete information. Imperfect information merely refers to the fact that some player's information sets in the game's extensive form are not all singletons. Incomplete information, on the other hand, refers to the fact that players' knowledge about the others' strategy choices (or preferences) is less than complete. This is clearly the case with real world markets. Harsanyi's analysis of such games treats players as Bayesians. Elsewhere I have considered an alternative Bayesian approach to incomplete games which is more suited to the analysis of imperfect markets than is Harsanyi's. The key difference is that in my model, players learn by sequentially revising their Bayesian prior distributions in order to reflect their experience in the game.

Ultimately, the sequence of revised distributions converges to a limiting endogenous objective distribution. The consideration of an endogenous objective distribution appears to be absent in Harsanyi's model.

Learning is a very important feature of economic systems. Market participants choose what they believe at the time to be their 'best' trade plans. Yet they are prepared to alter these plans as they learn about the market. With no auctioneer and incomplete information, a trader might choose prices that do not lead to market clearing: supply and demand might differ for some goods at various times. The game is not in a Nash equilibrium of the type described above. Typically, such markets experience 'adjustment' as the participants learn about its features. Ultimately, the key question of interest is whether the market tends to stabilize over time? By applying the techniques developed in Shefrin [4,5] I am able to show that joint learning by all market participants leads to market stabilization.

In the body of the paper, I construct the extensive form of a market game characterized by both imperfect and incomplete information. I discuss how players plan to behave in this game and describe what might be a reasonable definition of a noncooperative equilibrium. Briefly, the definition of noncooperative equilibrium which seems appropriate to this analysis is based on the notion that all players' learning processes have converged. Every player then acts as if he knows the above mentioned underlying objective distribution. This equilibrium is called an *informational equilibrium* and it is in the spirit of what economists call rational expectations.

§1. THE MARKET

For the purpose of illustration, it is useful to focus on an agricultural market. This merely serves as an example; the analysis can easily be generalized. To avoid later confusion, I henceforth refer to the participants in this market as players.

Every player can take one of the following three *types* of decisions at every point in time: production, consumption, or trade. Assume that a single player produces and sells only one type of crop; however he is interested in purchasing and consuming many crops. Crops are considered to be storable for some maximum period, after which time they perish.

A typical player faces the following situation. He has in his possession a certain quantity of crops that he has produced or purchased in the past. At any point in time, he must decide whether he should engage in further production, consume some or all of his accumulated stores, or engage in trade on the market. The first two activities are straightforward, though crop yield arising out of production is assumed to be randomly determined by nature.

Assume that trade takes place bilaterally between players at trading posts. Every trading post is a stationary location which is assigned to one and only one player. If a player is in the market and decides to remain in one fixed spot, he stations himself at his trading post. Other (mobile) players can then trade with him there.

Once in the market, every player announces a nominal price for the crop he sells. There may be many players at a given trading post (although only one is stationary). All players at the same trading post 'hear' each others' prices, although trade is only assumed to take place bilaterally between the stationary player and the mobile players at his post. On hearing the announced prices, the stationary player ranks the mobile players at his post from best to worst and proceeds to trade with the best. If there is more than one 'best' choice, one is chosen at random (with equiprobability). A specific trade occurs in the following way. Suppose the stationary player is selling corn at a unit price of 6 while a mobile player is selling wheat at a unit price of 3. The relative price of corn in terms of wheat is therefore 2. As an example, let the stationary player offer 10 units of corn for 20 units of wheat. The mobile player might offer 20 units of wheat for 10 units of corn and a mutually agreed trade then takes place. However, the mobile player might instead offer 10 units of wheat for 5 units of corn. Since this is a smaller trade offer than the one above, it is the only one which can be carried out. In other words, the short side of the market prevails.

A player's decision possibilities at a given point in time can be described in terms of a tree. See Figure 1. The tree is pretty much self-explanatory except perhaps for the term trade plan. A player's trade plan specifies the quantities he is willing to trade at any announced set of prices at the trading post. For example, the stationary player offered 10 units of corn for 20 units of wheat when the relative price of corn is 2.

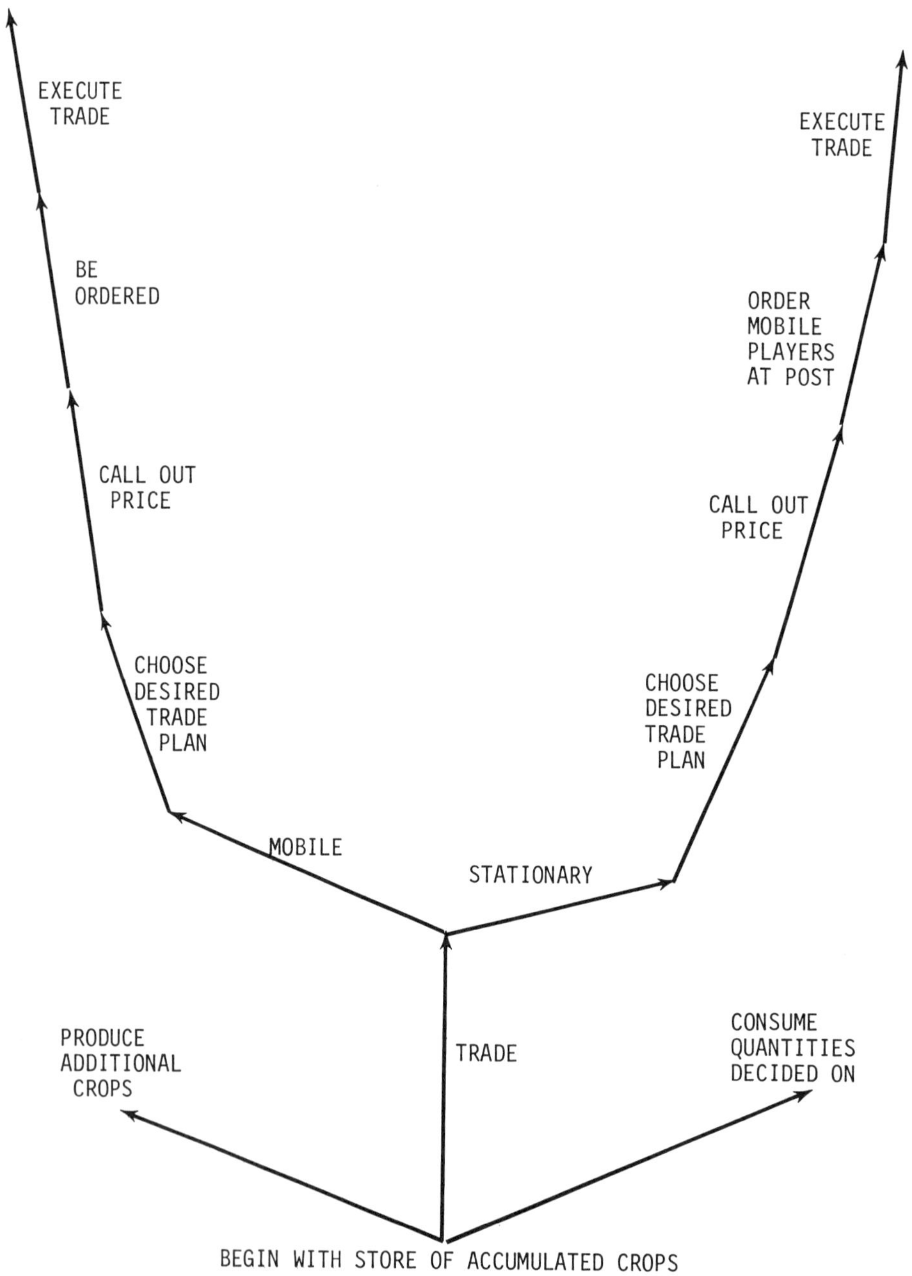
EXECUTE
TRADE
BE
ORDERED
CALL OUT
PRICE
CHOOSE
DESIRED
TRADE
PLAN
MOBILE
STATIONARY
EXECUTE
TRADE
ORDER
MOBILE
PLAYERS
AT POST
CALL OUT
PRICE
CHOOSE
DESIRED
TRADE
PLAN
PRODUCE
ADDITIONAL
CROPS
TRADE
CONSUME
QUANTITIES
DECIDED ON
BEGIN WITH STORE OF ACCUMULATED CROPS

FIGURE 1

§2. THE GAME; EXTENSIVE FORM

Assume that crops only come in indivisible quantities so that the decision possibilities are finite for every player. The dynamics of the market can easily be described in terms of the extensive form of a game.

Let there be N players and nature. Order the players 1, 2, ..., N. Assume that time is divided into periods and that the order of decision making is as follows. First, nature chooses a decision, and then every player sequentially chooses his decision according to the above order. Nature's only role is to determine the crop yield of any player who decides to produce.

The information sets describe every player's information possibilities. A player is not assumed to know what the other players and nature have decided to do in the current period when he himself must choose his own current decision. Once decisions have been made, however, a player learns about the prices quoted at the trading post at which he is located and the planned trade of the player with whom he trades at the relative price established. If he produces, then he learns the size of his crop yield. This information can be retained in memory. Assume that a player recalls all of his past experiences.

Each path in the game tree associates a consumption sequence to every player. Let c^i_{kt} be the quantity of crop k that player i consumes during period t. Let $c^i_t = [c^i_{kt}]$. Player i is assumed to have a current utility function $u_i(c^i_t)$ specifying the utility level incurred during period t.

The utility of the entire consumption sequence c_1^i, c_2^i, ..., is taken to be:

$$\sum_{t=1}^{\infty} \alpha_i^t \cdot u_i(c_t^i) < \infty \tag{1}$$

$0 < \alpha_i \leq 1$ is player i's discount factor. Notice that $c_{tk}^i = 0$ whenever i chooses to produce or to trade.

Player i's preferences are assumed to be von Neumann-Morgenstern. The utility function has been specified below, however the probability distribution is quite another matter.

§3. LEARNING

In the standard Nash-type noncooperative game, a player's probability distribution is determined by the other players' mixed strategies and the probabilities characterizing nature's behavior. Every player is assumed to choose a strategy which maximizes the value of his expected utility against these known distributions. Such knowledge is unrealistic for the above market setting. Every player would have to know the identity of all the other players (including their information sets and preferences) as well as their strategies (consumption, production, and trade plan). However, competitive markets are on the whole quite impersonal. Trade mostly takes place on the basis of prices, and prices are by and large subject to considerable uncertainty. In this setting, incomplete knowledge of the other players' trade plans is the rule rather than the exception.

If a player does not maximize against the other players' and nature's known mixed strategies, against what does he maximize? How does he view the behavior of the market? In a large competitive market, it seems reasonable to assume that a player views the behavior of the rest of the system in the same way he views nature. By this I mean that he considers his own role to be so small as to make strategic analysis not worthwhile. Instead, the player assumes that the market behaves as if its actions were generated by an underlying objective stationary stochastic structure. Consequently, he would want to act like a dynamic programmer involved in a Markov decision process, and compute his strategy as an optimal dynamic programming policy. This is certainly consistent with most economic analyses of individual behavior in competitive markets.

Need we assume that a player knows this stochastic structure from the outset? Not at all! Indeed the analysis is most interesting when such knowledge is initially unavailable. For then, the player must formulate a policy (strategy) in which he both learns about the behavior of the market from his own experience and observation, and simultaneously tries to do the 'best' for himself. Learning can take place on many levels. First, a player can learn about the general structure of the market (location of trading posts, and so on). Second, he can learn about the type of information he receives. In some cases, a player may encounter a situation which he previously did not even consider to be within the realm of possibility. O'Driscoll [3] has referred to this problem as "incomplete listability" and has criticized rational expectations economic models because it fails to take this feature into account. I shall comment on this below. Third, a player can learn about his own strategy set as he interacts with other players in the game. Finally, he can learn about the relative frequencies with which repeating observable events occur over time. For now, I confine my attention to this last aspect of learning, it being the most difficult. Toward the end of the paper I shall discuss how the basic framework can be suitably modified to take account of the other features.

Consider what happens when a player with little or no previous experience begins to make decisions. With what pieces of knowledge might he *not* be acquainted? For example, would he know the probability distribution governing his crop yield? Would he be familiar with the prices that the other players call out both at his own and others' trading posts? Would he be aware of how likely it is that he will be able to carry out his planned trade when he announces the price he is charging for his crop? In a Nash equilibrium, the answer to all these questions is, of course, yes; however the answers are most probably no for the inexperienced trader. A player whose knowledge about these matters is incomplete would certainly want to use his experience in the game to learn the answers. For example, he would observe and record the relative frequencies of his crop yield in order to ascertain that aspect of nature's strategy. Of course, the player implicitly assumes that nature's choice is governed by a stable probability distribution. If the total number of players is large, it seems reasonable that he will also treat the market in much the same way as he treats nature. That is, he will assume that the market acts as if its behavior was generated by an underlying probability distribution. Thus, the player would record the frequency with which other players call out various prices, he would keep track of the way market offers come in response to his calling out various prices, and so on.

When a player bases his estimate of the probability of some event on relative frequencies, he is obviously treating the repetition of that event as being governed by the same stochastic forces at each occurrence. Now an event can be repeated if it is based on the finite past. For this reason, I assume that all players consider those events which occurred more than some finite number τ periods ago (relative to the current period) as being irrelevant to the determination of present events. It seems reasonable to take τ to be the maximum amount of time that crops can be stored without perishing. Therefore the information that a player retains about events which occurred more than τ periods ago is only used as historical data for the computation of relative frequencies.

Consider a player who proceeds to do the 'best' he can for himself as he learns about the market. Under what type of conditions might he exhibit something resembling Nash behavior? Suppose, for instance, that his learning process begins to converge over time. As he observes events that repeat themselves over and over again, he comes to understand the underlying structure of the market as he perceives it. Previously, his beliefs about the relative frequencies of these events were changing. As a result, he most probably altered the decision he made in response to some observable event from time to time. In other words, he may have taken one particular decision the first time the event occurred, but taken some other decision the second time, and so on. This is because he learned something about the pertinent probabilities in the interim. Once he ceases to learn, he begins to react to the same situation in the same way each time it occurs. From this time forward, he maximizes the value of his expected utility against what he believes to be the true stochastic structure. This is because he receives no disconfirming evidence to the contrary. This is quite close in spirit to the Nash concept where a player maximizes his expected utility against the correctly perceived strategies of the others. However, in the present framework, a player need not even understand the structure of the game tree, let alone know the strategy of every other player.

If all players simultaneously find themselves in a position where they cease to learn about the game, then the game is in an equilibrium of a sort. This is because the information that each player receives leads him to repeat his behavior. Therefore I call this equilibrium an *informational* equilibrium.

During the time a player is learning, he may find that he cannot carry out his intended trade plan. For example, there may be insufficient supply to meet his demand at the price he announces. One reason for this might be that he calls out "too high" a price for his own crop. If it is too high, he would ultimately learn this fact over time by noting that he consistently fails to carry out his trade plan. However, I hasten to add that he may still find that he cannot carry out his intended trade plan even when the game is in informational equilibrium. The difference is that the latter situation is intermittent, while the former is persistent. The latter may occur because, for example, some other player has had a poor crop yield and our player does not have information about this fact. Consequently, he sets his own price according to the information he has. (If he knew about the other player's crop yield he might do something different.) Therefore, his price is correct 'on average,' though it may lead to supply-demand imbalance from time to time. What must be understood is that in informational equilibrium the frequency with which this temporary imbalance occurs is foreseen and taken into account.

§4. BAYESIAN LEARNING

In order to pursue the question of learning in more depth, I shall have to introduce some formal notation. Let D_i denote the set of possible decisions for player i. Recall that a strategy specifies a decision for player i at each of his decision nodes. D_i is assumed to be finite. This implies that crops come in indivisible units and that the crop quantity is bounded at every moment. The possible prices that player i can quote are also finite.

Every decision node for i in the game tree belongs to some information set. Let μ_i^t be an information set for period t. μ_i^t describes the past history encountered by i. Assume that i has perfect memory. I denote the event described by the most recent τ periods as β_i^t. Clearly β_i^t can be repeated over and over again. Thus, if $t1 > t2$, then μ_i^{t1} and μ_i^{t2} can describe the same event $\beta_i^{t1} = \beta_i^{t2}$, even though $\mu_i^{t1} \neq \mu_i^{t2}$. The crucial assumption in the analysis is that i believes that the next event β_i^{t+1} he will observe is determined by a conditional probability distribution $\text{Prob}\{\beta_i^{t+1}|\beta_i^t,d_i^t\}$,[1] (where i observes β_i^t in period t and chooses d_i^t. Initially, he thinks that the market is governed by such a distribution; however, he does not know what the distribution is. Furthermore, the distribution is stationary so that the t and t+1 superscripts are superfluous. However, he is assumed to have initial beliefs about this distribution which are represented by a Bayesian prior function $\pi_0^i(\cdot|\beta_i,d_i)$ (time subscripts have been omitted). $\pi_0^i(p|\beta_i,d_i)$ is player i's initial subjective probability density that $\text{Prob}\{\cdot|\beta_i^t,d_i^t\} = p$, where p denotes an arbitrary probability distribution on the set $\{\beta_i^{t+1}\}$.[2] For now, assume that $\pi_0^i(p|\beta_i,d_i)$ is strictly positive for almost all p.

Suppose that player i confines his attention to those periods in which he observes β_i and chooses decision d_i. Index those periods by τ_j. Then player i can form the sequence of posterior distributions

$$\{\pi_{\tau_j}^i(\cdot|\beta_i,d_i)\}$$

by using Bayes' Rule. Let β_i^t denote the event that i observes in period t, $p(\beta_i^t)$ denote $\text{Prob}\{\beta_i^t|\beta_i^{t-1},d_i^{t-1}\}$ when the latter equals p, $\beta_i^{t-1} = \beta_i, d_i^{t-1} = d_i$, and ν is of the same form as p. Define[3]

---

[1]For example, the amount of crops available to player i depends on his accumulated stocks and consumption during the recent past.

[2]p belongs to a unit simplex of appropriate dimension.

[3]The range of integration in (2) is the simplex mentioned in footnote 2.

$$\pi^i_{\tau_{j+1}}(p|\beta_i,d_i) = \frac{p(\beta_i^{\tau_j}) \cdot \pi^i_{\tau_j}(p)}{\int \nu(\beta_i^{\tau_j}) \cdot \pi^i_{\tau_j}(\nu)d\nu} \tag{2}$$

(τ_j = t in the above expression.)

As is well-known, i's subjective estimate of the unknown $\mathrm{Prob}\{\cdot|\beta_i,d_i\}$ is just[4]

$$\int p\pi^i_{\tau_j}(p)dp \tag{3}$$

Since an information set μ_i^t exhibits perfect memory, an integral of the form (3) can be associated with every (β_i,d_i) and $\tau_j < t$. From this point on, consider only that integral which is based on the $\pi^i_{\tau_j}$ with the highest τ_j (for each (β_i,d_i)).

Player i's strategy specifies the decision d_i that he takes at every one of his decision nodes. He chooses an 'optimal' policy by forming an associated Markov decision process and solving the associated infinite time dynamic programming problem. A state of this process is simply a representative information set μ_i^t. The transition probabilities are given by (3) (see the above remark) and the utility function is given by (1). An optimal policy is a function δ_i which maps an information set μ_i^t onto a decision $d_i^t = \delta_i(\mu_i^t)$. Observe that player i considers the degree to which he expects to learn and takes this into account through (3) when he chooses his strategy (policy) δ_i. In particular, this means

---

[4]The range of integration in (3) is the simplex mentioned in footnote 2.

that we can speak of his doing the 'best' for himself while he learns in a well defined and unambiguous way.

§5. EQUILIBRIUM

A player ceases to learn when for each (β_i, d_i) his sequence of posteriors $\{\pi^i_{\tau_j}(\cdot|\beta_i, d_i)\}$ begin to converge and mass on one particular distribution $\bar{p}$ as $\tau_j \to \infty$. In this case, we say he has learned that $\text{Prob}\{\cdot|\beta_i, d_i\} = \bar{p}$. If he never observes β_i, or never chooses d_i when he observes β_i, then τ_j cannot go to infinity. In this case he does not learn $\bar{p}$, but thinks it unimportant or not worth finding out about.

Suppose that we construct a Markov decision process whose state space is $\{\beta_i\}$, not $\{\mu^t_i\}$, and whose transition probability is given by the set $\{\bar{p}\}$ above (or (3) when $\tau_j \nrightarrow \infty$) instead of always by (3) as before. In this case, we obtain an optimal policy $\tilde{\delta}_i$ which maps each event β_i into a decision $d_i = \tilde{\delta}_i(\beta_i)$. Recall that β_i is the portion of μ^t_i that is relevant for decision making (apart from the historical data for the estimation of transition probabilities which is now embodied in $\{\bar{p}\}$). Clearly $\tilde{\delta}_i$ describes that portion of player i's strategy δ_i which is relevant once his learning process has converged. Notice that i chooses decision $d_i = \tilde{\delta}_i(\beta_i)$ every time he observes the event β_i to occur. If all players' learning processes converge in this way, the game is clearly in informational equilibrium.

The main result from Shefrin [5] which I want to apply here is essentially this. Let nature's strategy be chosen according to a stochastic process which is Markovian of order τ or less. Consider any period $t < \tau$ and let player i's

Bayesian prior function $\pi_t^i(\cdot|\beta_i,d_i)$ be positive almost everywhere on its respective domain. Then if all players choose expected utility maximizing strategies, (the mass of) $\pi_t^i(\cdot|\beta_i,d_i)$ converges onto an endogenous objective distribution with probability one for all i.

§6. ADDITIONAL ISSUES

The above result is easily interpreted for our agricultural market. If players have complete knowledge of the possible outcomes, but incomplete knowledge of the relative frequencies with which these outcomes occur (as a result of their choosing various decisions), then Bayesian revision will ultimately lead to these frequencies becoming known. The individual player sees himself as eventually learning how the market objectively behaves. For example, he learns the probability distribution of his crop yield, the frequency with which various prices are called out in the market, how often he fails to carry out his intended plans, and so on. Things are more complex from the system point of view. Initially, players may make correct subjective but incorrect objective decisions (as is to be expected since learning is taking place). Consequently, the market may experience a certain amount of adjustment as players alter the way they respond to given circumstances during the learning process. Ultimately, though, the system stabilizes and goes into informational equilibrium (by the above result).

What exactly is meant by stabilization? Does it mean that a single price prevails for the same good over time from some period on? Or does it mean that markets always clear?

The answer is no to both. After all, crop yields vary according to nature's strategy, and information differences among players may prevent market clearing. For example, a given player's intended trading partner may incur a bad crop yield and decide to continue production rather than coming to market. However, the given player may be unaware of this fact, thinking instead that his prospective trading partner is elsewhere in the market (that is, at another trading post).

By stabilization, I simply mean that the market reacts in the identical way to identical circumstances so that some form of stochastic stability prevails.

Clearly, this game takes place in two stages. In the first stage, the game is not in equilibrium: players are learning. In the second stage the game is in informational equilibrium: all players' learning processes have converged. The above quoted result considers only one type of learning--learning about relative frequencies. Yet, as I mentioned earlier, several types of learning can be distinguished. One of these is "incomplete listability," mentioned by O'Driscoll in his critique of rational expectations. Incomplete listability refers to the fact that a player may be unaware of all possible outcomes. For example, some player may be unaware of the existence of a particular crop. Notice that "complete listability" is actually assumed within the hypothesis of the convergence result quoted in the previous section. Yet, "incomplete listability" can be easily introduced into the basic framework. Suppose, for example, that a player initially fails to foresee all possible outcomes. He may be aware that he is ignorant in this regard, or then again he may not be: it turns out not to matter. Suppose he observes a 'surprise' event which he had never even considered before. What ought he to do? As is well-known, he cannot use Bayes' Rule properly in this case. Obviously, the sensible thing for him to do is to 'reprogram' himself and modify the domain of his Bayesian prior to include the new outcome. How he does this is not important for present purposes so long as the new Bayesian prior attaches positive probability to the reoccurrence of the new outcome. Clearly he ought to do this every time he encounters a 'new' outcome. As model builders ourselves, however, there are no surprise outcomes. We know the set of all possible outcomes and can therefore construct a theory which takes all this into account in a well-defined way. In view of the inherent assumptions on finiteness, any player whose knowledge is initially characterized by "incomplete listability" will ultimately learn about all the encounterable surprise outcomes with probability one. Therefore the game can actually be thought of as occurring in three stages. First, players learn about all the surprise outcomes that they will ever encounter (though, of course, there may be others that they never observe). At the end of this stage, all the hypotheses of the convergence result are satisfied and the next two stages are, respectively, learning (as before) and informational equilibrium. Clearly, learning about new outcomes is but one additional type of learning, although the same method can be used for other types. Notice that rational expectations refers to the third stage only which is why it fails to take almost all learning into account (including incomplete listability).

There is one other aspect of the model which is particularly interesting. In standard Bayesian convergence theory, the player is involved in a game with nature alone and is therefore learning about a stationary objective distribution. However, in an N-person game, all players are simultaneously learning. Therefore each is altering his reactions to repeating circumstances as he learns. As a result, there is no underlying objective distribution during the learning phase. Nevertheless, so long as players believe that such a distribution exists (and is based on the most recent τ periods), these beliefs will ultimately turn out to be correct.

Clearly the learning phase is extremely important in shaping the ultimate direction the system takes. For example, a player may have had a bad experience when he took a particular decision early on. This may lead him to avoid taking that decision forever after. Now it may turn out to be the case that his bad experience would never be repeated in equilibrium if he repeats the decision: perhaps it only resulted because some other player was also learning at the time and made some mistake. Nevertheless, the mistake gets carried through into equilibrium. Now an astute player who understands this feature may try to achieve the same effect intentionally! That is, he may choose the 'mistaken' decision in order to influence the others' future behavior even if he incurs some cost at the time. When the manipulating player would not act in this way throughout, his behavior is called *deceptive*. Deception constitutes an important deviant from competitive economic behavior and is usually manifest in cooperative situations (such as oligopolies, trade unions, etc.). As such, I do not deal with the issue here, except to draw attention to its existence.

REFERENCES

1. Hahn, F. H., *On the Notion of Equilibrium in Economics - An Inaugural Lecture*, Cambridge University Press, Cambridge, U.K., 1973.
2. Harsanyi, John C., Games with Incomplete Information Played by 'Bayesian' Players, *Management Science*, Parts I-III, *Management Science*, *14*, 159-182, 320-334, 486-502 (1967-68).
3. O'Driscoll, Gerald, Rational Expectations, Politics, and Stagflation. In *Time, Uncertainty, and Disequilibrium: Exploration of Austrian Themes*, Mario J. Rizzo, ed., Lexington (forthcoming).
4. Shefrin, H. M., Games with Self-Generating Distributions, mimeo, 1977.
5. Shefrin, H. M., Convergence of Joint Learning Processes in Infinite Horizon Games, mimeo, 1977.